DATE DUE			
OCT 3 0 1973			
Jul 27 83			

EVOLUTION OF LIFE

*'Happy the man whose lot is to know
the secrets of the earth. He hastens not
To work his fellow's hurt by unjust deeds
But with rapt admiration contemplates
Immortal Nature's ageless harmony
And when and how her order came to be.'*

Euripides

Evolution is the phenomenon of perpetual, yet always harmoniously balanced change. From minute to minute, the very earth, its plants and its animals, are never the same. Over even greater periods of time the changes are truly tremendous. Continents rise and fall. Lifeless lands become the habitat of venturesome fish. Reptiles grow feathers and fly. Yet all these seemingly fantastic developments occur in an orderly way. Organisms change as part of the overall environment and everything that exists at a given moment in time relates to everything else.

This book covers the main evolutionary events that have occurred over the long span of earth history. The evidence is examined and assessed, and the intricate and interrelated mechanisms of heredity and environment that have led to the present are discussed.

G.Z.

A GROSSET ALL-COLOR GUIDE

EVOLUTION OF LIFE

BY CATHERINE JARMAN
Illustrated by Peter Thornley

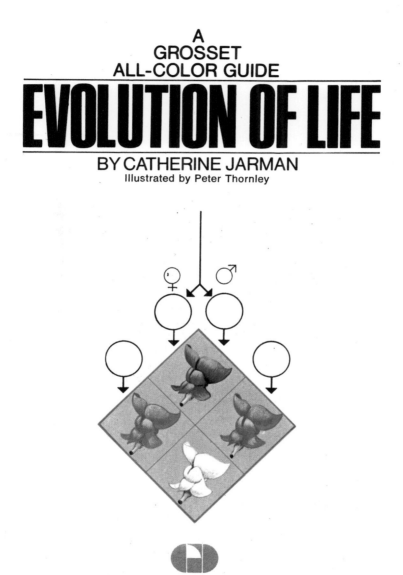

GROSSET & DUNLAP
A NATIONAL GENERAL COMPANY
Publishers ▪ New York

575
J29e
113511
ajn.1980

CONTENTS

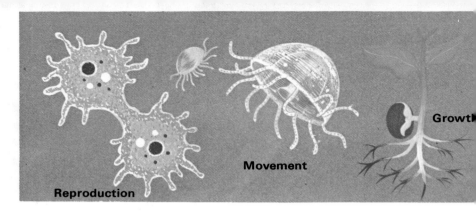

Reproduction

Movement

Growth

INTRODUCTION

The nature of life

Most of us know what living things are. We have little trouble distinguishing living from nonliving forms, yet a precise definition of 'life' is very difficult. About the best we can do is to list some of the characteristics which all living things have in common.

The most familiar characteristics of living things are growth, movement, metabolism, reproduction, irritability, ingestion, respiration and excretion. The basic chemical composition of most living organisms consists primarily of water, in which a number of organic compounds and salts are mixed. The differences in animal life lie not in the basic chemical elements but in slight variations in the arrangement of these elements and molecules.

Another property basic to life, in spite of much scientific work, is that all life comes from some previously existing life. This endless chain of life and the changes that have been brought about through time and the natural forces is the story of evolution.

Common processes of life (*above*) and diversity of forms (*below*)

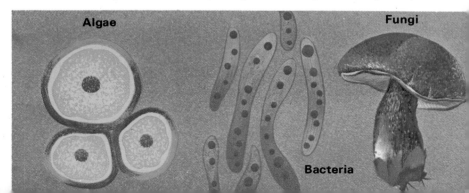

Algae

Fungi

Bacteria

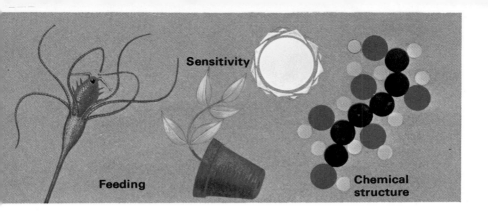

Sensitivity

Feeding

Chemical structure

The diversity of life

From the few organisms — perhaps only one — in the Precambrian, life has persisted in an unbroken line to the present day. That it persisted at all is a marvel in itself, but it also managed to become diversified into an array of different forms. The great diversity of living forms that exists today is shown by the large number of species in certain groups:

Angiosperms (flowering plants)	200,000
Thallophytes (algae, fungi, lichens)	107,000
Bryophytes (liverworts, mosses)	23,000
Pteridophytes (ferns, horsetails)	10,000
Invertebrates (insects, worms)	1,084,400
Vertebrates	42,600
Fish	25,000
Amphibians and reptiles	5,500
Birds	8,600
Mammals	3,500

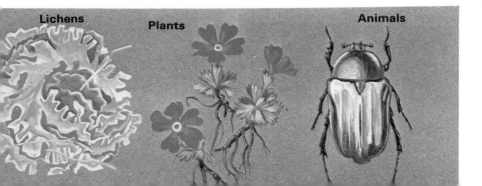

Lichens

Plants

Animals

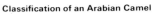

Kingdom — Animal

Phylum — Chordata

Class — Mammalia

Order — Artiodactyla

Family — Camelidae

Genus — *Camelus*

6

Species — *dromedarius*

Classification

For centuries man has tried to arrange living organisms into some kind of logical order or grouping. This arranging in scientific groupings is called classification or taxonomy. However, it is nearly the art of the impossible, as perfect classification is nonexistent. Nevertheless, biologists strive for perfection for, after all, animals and plants cannot be studied fully until they have been classified.

At first, animals were grouped according to how they were adapted for living. Thus, whales were classed as fish solely because they lived in the sea. Similarly, bats were grouped as birds only because they could fly. This classification based on the adaptation of the animal led to many wrong groupings and did not show the relationship between the various animals.

The next phase was an improvement but was still inaccurate by modern standards. This grouping, known as morphology, was based on the form of the animal. This system was best set forth by Linnaeus (1707–78), a Swedish naturalist. It is based on the assumption that there are as many types of animals now in existence as were produced during the six days of creation. His thoughts followed the traditional assumptions of that time about the kinds of living things. Linnaeus was well aware of the anatomical resemblances between certain organisms, for example, horse and donkey, newt and salamander, ape and man. He correctly grouped lions, tigers and domestic cats into a single group called the Felidae. These animals, although they exhibited many resemblances, were merely grouped according to the overall structural plan. All species, it was assumed, remained unaltered through the course of time and a possible link between generations was not realized. This form of classification, and the assumptions which led to and from it, did not suggest evolution.

When the fact of evolution had been accepted in the second half of the 19th century, taxonomists tried to express the evolutionary relationships between animals and plants. The terminology used by Linnaeus was retained, but a wider range of evidence was taken into account. Morphology was still very important, but the evidence of fossils (paleontology), genetics, embryology, immunology and geographical distribution was also taken into consideration.

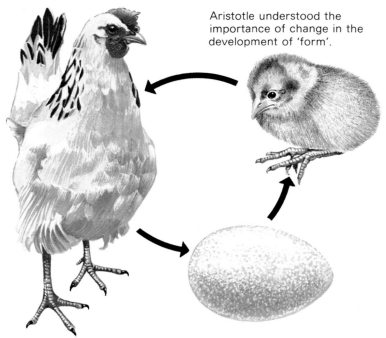

Aristotle understood the importance of change in the development of 'form'.

EVOLUTION

Some early theories

Today, evolution is accepted as a fact by the majority of people. However, it has not always been a scientific theory. After all, the notion of life evolving is not one which comes immediately and naturally to the majority of minds.

Many ideas of evolution were presented in symbolic forms in myths. Thales (640–546 B.C.) believed all life came from water, whereas Anaximenes of Miletus (circa 500 B.C.) said all things came from air. Most mythological ideas are linked with the 'elements' such as earth, fire, water, and air. Some ideas were quite fantastic and comparatively recent. A. Kircher (1601–80) thought orchids gave birth to birds and small men; B. de Maillet (1656–1738) though recognizing the true nature of fossils, thought birds were derived from flying fish, lions from sea lions, and men from mermen.

Theological thoughts interpreted the idea of evolution as an act of God. The Hebrews accepted the religious writings of the

Bible, believing that living things were especially created in six days by an all-powerful personal god. Special creation was an act that was apparently done once and for all. Although the biblical account gives six 'days', this ancient term is interpreted to mean six epochs. The special creation theory is based on the analogy of the making of the universe by God, with the manufacture of articles by man. However theological views and the evolution theory are compatible, if we consider that both have a common background in the production of life.

The theories of the Greek 'thinkers' give the other main thoughts on evolutionary ideas. In his biological philosophy, Aristotle (384–322 B.C.) showed the importance of change and the theory of 'forms' or 'essences'. He understood that the speck of matter in a hen's egg becomes, by definite stages, a chicken. Also, by numerous examples he shows that he understood complete metamorphosis where organisms such as frogs and insects, developed and matured in definite changing stages. His work also shows he conceived the idea that life began as undifferentiated 'matter' with a 'potentiality' for turning into a form although his ideas were limited to individuals. Any

Special creation in six days is described in the Bible.

evolutionary thought of how one individual can somehow change into another was absent.

The act of special creation and the Aristotle philosophy of 'forms' were quite sufficient to satisfy nearly everybody during the Middle Ages and even lasted through the Renaissance and Reformation, when one might have expected some fresh thoughts on the philosophy of biology.

Up to the 18th century the Earth was assumed to have been created in 4,004 B.C., as calculated by the Archbishop of Armagh (1581–1656). It was not until the science of geology rapidly advanced in the late 18th century that doubt was thrown on the 6,000-year-old Earth theory. Several scientists doubted the few-thousand-year theory and suggested several hundred thousand years as more appropriate.

During the late 17th and into the 18th century a true appreciation of fossils developed. Up to this period fossils were thought to develop from moist seed-bearing vapors, blown from the seas into the crevices of the earth. Martin Lister (1638–1712) and Robert Hooke (1635–1703) suggested the possible value of distinctive fossils in relation to the strata in which they were found. Lister did not, however, think they were fossilized animal or plant remains.

No startling theories followed until 1788, when James Hutton (1726–97), an Edinburgh doctor, published an essay

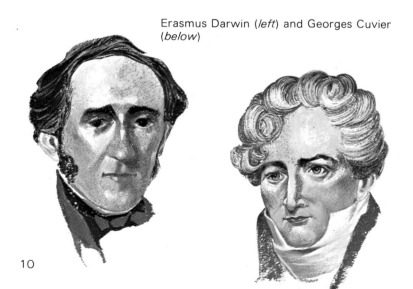

Erasmus Darwin (*left*) and Georges Cuvier (*below*)

entitled 'A Theory of the Earth' (1795). 'Uniformitarianism' was Hutton's explanation of the past history of the Earth. By this, Hutton stated that divinely ordained floods and universal rock upheavals were unnecessary to explain the existence and nature of oceans and mountain chains. Hutton argued that the Earth's processes were gradual in their nature, rock and soil being worn away by the process of erosion. His ideas led him to state that the continents could not have been formed in a few thousand years, but took millions of years. He implied that the Earth was infintely old and had an equally vast future.

Although the concept of the alteration of the Earth was accepted, the changing of animals and plants was still not suggested. At the turn of the 18th century, however, the work of William Smith (1769–1838) did a great deal in establishing the value of animal fossils in the recognition of strata in geology.

Georges Cuvier, a zoology professor in Paris, and Smith's contemporary, had a profound knowledge of comparative anatomy which enabled him to reconstruct animal skeletons from fossil bones or markings. This added greatly to the knowledge of past animal life.

This great spurt in the geological sciences provided a wealth of information about animal and plant characteristics. Many species of plants and animals described by geologists were extinct, so the idea of a succession of animal types through time began to unfold. However, no one crossed the threshold of an evolutionary interpretation of life. One fact that helped to prevent this was the non-evolutionary interpretation of animal life provided by Cuvier himself — the theory of 'catastrophism'. He concluded that each fauna had been destroyed by a gigantic catastrophe and a new fauna then migrated from distant regions. Noah's flood was taken to be the most recent catastrophe.

Erasmus Darwin (1731–1802) is perhaps most famous for being grandfather to Charles Darwin. However, his own thoughts and speculations caused a sensation in his day, both in England and on the continent. His notable scientific work was 'Zoonomia', an attempt to find out the organic laws of life. His natural philosophy was that spirit and matter are the foundations of nature. He defined life as being due to a special force, irritability, and all manifestations of life result from the contraction of fibers, induced by irritation.

Lamarck (*left*) and the classic example for his theory of evolution (*right*)

The theories of Lamarck

Jean Lamarck (1744–1829), a friend of Cuvier, was the first person to tackle the problem of how species originated. Boldly, he proclaimed that there was no essential difference between species and varieties, that species, like varieties, were subject to change, and that 'transformation', not immutability, was the basis of life.

In his work, 'Zoological Philosophy' (1809), Lamarck suggested that a transformation of species might have occurred by the 'inheritance of acquired characters'. He drew attention to the observation that animals do change in form during their life span. For instance, a man may build up his muscles during his life by repeated exercise. Assuming this acquired character was passed on or inherited by the offspring, then the next generation would be born with slightly stronger muscles. In his 'Zoological Philosophy', Lamarck argued that if this change was continued through several generations, it would result in a completely new species.

The giraffe is the classic Lamarckian example. The giraffe with its long neck could have developed from an animal with a much shorter neck. Straining its neck through untold generations to reach higher levels of foliage presumably resulted in an elongated neck. This was the basis of Lamarck's theory, that species gradually changed in form by the use or disuse of any particular organ or faculty.

Although Lamarck was wrong about the mechanisms of evolution, he did introduce the idea that species are not immutable. Lamarck did no experiments to provide evidence to support his theory at the time, and a considerable amount of later experimenting failed to produce examples which convincingly showed any inheritance.

Charles Darwin (1809–82)

Darwin—his life and theories

The 'theory of evolution' is usually associated with Charles Darwin, whose genius in the late 1850's gave the world the new idea of natural selection.

Darwin was born in Shrewsbury, England, in 1809, the son of a wealthy doctor. At the local public school he received the usual course of Latin and Greek verses with classical geography and history. However, he became more interested in natural history and was subsequently sent to Edinburgh to study for the medical profession. The sight of two operations under the conditions of that time, before anesthetics had been brought into use, revolted the 17-year-old Darwin. After two years, when it was quite apparent he was totally unsuited for medical life, he returned home to Shrewsbury. His family now decided he should study for the Church. Charles did not violently object and so 1828 found him in Cambridge. On graduating, Darwin was resigned to take Church Orders when fate intervened to set an apparently unambitious man, with little distinction to his name, on the way to show the world his spark of genius and depth of reasoning. On the advice of his professor and friend, John Henslow, Darwin joined the survey ship *H.M.S. Beagle* to journey to South America as an unpaid naturalist.

thus, 'natural selection' would take place. Although Darwin had the framework to his evolutionary theory in 1838, it was another 20 years before the public was to learn of his revolutionary ideas, when his hand was forced by Alfred Russel Wallace.

Wallace had read Malthus in the early 1840's, but it was not until 1847 that he too was convinced of evolution. In that year he joined a collecting expedition to South America and later worked in Malaya and the eastern Indian archipelago. Surrounded by evolutionary evidence, Wallace wrote a short paper arguing that species must have come into existence slowly but he, like Darwin, was worried about how the change took place. In 1858, while recovering from an attack of malaria, Wallace thought of Malthus and suddenly came to the same conclusion as Darwin. When recovered from his illness, Wallace quickly wrote a 20-page essay outlining the theory. He then sent this identical theory to Darwin in Kent to ask his opinion for publication.

Seeing his own thoughts written by a distant, unknown person must have shocked Darwin. He realized he had no time to lose, and on the advice of his friends wrote a short paper stating his views and had his paper along with Wallace's read to learned naturalists at the Linnean Society in London in July 1858. Darwin followed this by publishing his book 'The Origin of Species' or 'The Preservation of the Favoured Races in the Struggle for Life' in November 1859. In his

Pouter

Short-faced tumbler

book Darwin put forward four points he knew to be true and three which we now know to be true. They can be briefly summarized as follows:

1. Plants and animals produce a far greater number of reproductive cells than ever give rise to mature individuals.
2. The number of individuals in a species remain more or less constant.
3. Because of this there must be a high mortality rate.
4. Members of the same species are not identical but show variety in many of their characteristics.
5. Because of these variants, some are more successful than others in the competition for survival, and these will be the parents of the next generation.
6. Hereditary resemblance between parent and offspring is a fact.
7. Subsequent generations will gradually maintain and improve on the characteristics which allow them to survive.

The first edition of the book was sold-out on the first day of publication. Although the religious Victorians were eager to read it, they were also shocked and disgusted. The religious issue and the implication of man's evolution from ape-like forms (carefully avoided by Darwin) became greatly disputed and argued.

A public debate on the issue occurred in 1860. The orthodox followers were represented by the Reverend Samuel Wilberforce, Bishop of Oxford; Darwin's theories were defended by Thomas Huxley, a trained medical surgeon. Although a bril-

Huxley (*left*) and Wilberforce (*below*)

By de-tailing mice for many generations, Weisman hoped to obtain evidence to discredit Darwin.

liant man, Wilberforce had no knowledge of the evolutionary issues and cared little for scientific philosophies. Huxley spoke in an effective way, pointing out the Bishop's faulty scientific attitude and explaining the new ideas in a precise and factual manner. However, the evolutionary conflict was not completely settled and in certain aspects it still exists today.

Weisman (1834–1914), a German biologist, was the first major critic of the Darwinian theory of evolution and Lamarckian theory. Between 1868 and 1876 he published a series of papers contending that acquired characteristics of all genetic or body variation could not be inherited. He made little use of practical evidence, although he had carried out a series of crude experiments such as de-tailing mice for many generations to show that a de-tailed mouse never produced a tailless mouse.

He assumed that there was a distinction between a body of an organism—the soma—and the cells concerned only with the reproduction—the germ plasm. He postulated that only the germ plasm could affect inheritance and that the soma could play no part. He also made biologists conscious of chromosomes by identifying them with the hitherto hypothetical particles within them. Weisman's theory was acknowledged and, although unacceptable today, in the 1880's it meant that Lamarckian views were discredited.

Table showing fossil-bearing geological periods

a circle indicates extinction

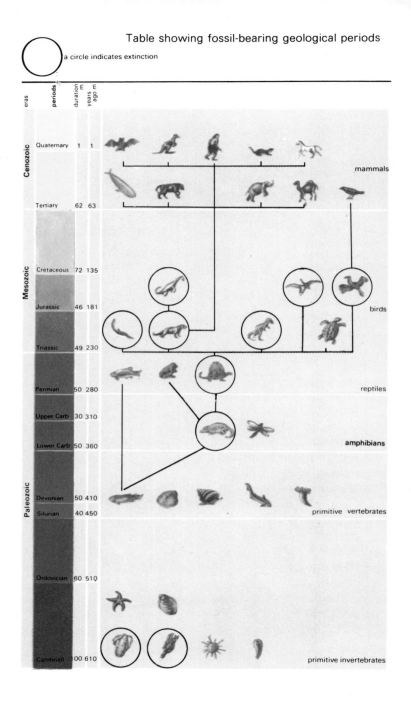

eras	periods	duration m years	ago m
Cenozoic	Quaternary	1	1
	Tertiary	62	63
Mesozoic	Cretaceous	72	135
	Jurassic	46	181
	Triassic	49	230
Paleozoic	Permian	50	280
	Upper Carb	30	310
	Lower Carb	50	360
	Devonian	50	410
	Silurian	40	450
	Ordovician	60	510
	Cambrian	100	610

mammals

birds

reptiles

amphibians

primitive vertebrates

primitive invertebrates

EVIDENCE OF EVOLUTION

Using methods based on the radioactive properties of uranium and thorium, modern estimates of the age of the Earth have been given as between four to five billion years. Studying fossils has helped to establish the order in which rocks have been laid down during the course of time. The geological time scale serves as a standard frame of reference for the description and correlation of various isolated events in the history of life on the Earth.

Geological history is broadly divided into eras, founded upon the general character of the life they represent. Thus, there are the Cryptozoic (hidden life), Paleozoic (ancient life), Mesozoic (medieval life) and Cenozoic (modern life) eras. Each era is subdivided into periods, usually based upon selected sequences of stratified rocks which display certain fossils.

Abundant fossil material does not appear until lower Cambrian times. This is a quite late stage of geological time (see pages 72 and 73). The pre-Cambrian era represents perhaps nine-tenths of the Earth's history.

Man
First mammal
First reptile

First fish

First life

Earth was formed

The geologic time scale

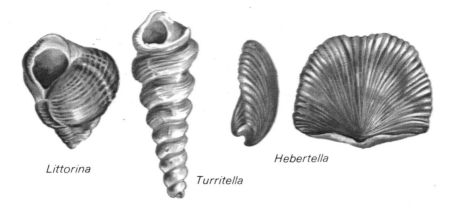

Littorina

Turritella

Hebertella

Millions of fossil animals and plants have been unearthed, all of which add to our knowledge of the diversity of life in the past. Fossil evidence helps to verify the ideas of evolution but, because of many gaps in the record and the impossibility of knowing the conditions under which these changes took place, evidence from other areas of knowledge is also needed.

Paleontology

Unaccountable trillions of animals and plants have lived and died on earth during its long existence. Of these, comparatively few died in a place where their form could be preserved. Nevertheless, there is an abundant wealth of fossil material which has been discovered, although unknown numbers still wait to be unearthed. Fossils are the remains or traces of any recognizable organic structures that have been preserved since prehistoric time.

By far the most abundant fossils are the pollens and spores and the shells of microscopic organisms which settled to the bottom of prehistoric seas. Those with hard parts are more suitable for preservation, and so shellfish such as oysters and lamp shells were often preserved along shore strand lines and in offshore water.

Preservation may take place in several ways. Petrifaction, widely used in reference to fossils, means 'turned to stone',

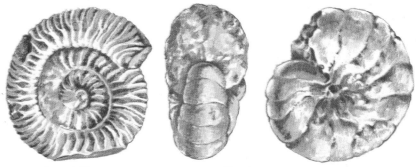

Agoniatites   Stephanoceros

A selection of fossils

although this very rarely happens to the whole of an animal. Huge fossil logs of the petrified forest of Arizona which became buried in the old Triassic flood plains are good examples. They became well fossilized due to the volcanic sedimentary material which covered them, and the silica of this material gradually replaced the wood of the trees.

Impressions are important fossil forms also. An animal becomes buried in fine-grain sediment, and a mold is made by the sediment of the animal's internal or external shape. These internal and external molds are negatives of the original. The two impressions found in the Solnhofen shales of Bavaria of a Jurassic bird *Archaeopteryx* were formed in this manner. Tracks and paths of animals appear as imprints or compressions. These fossils are frequently clearly displayed on prehistoric mud flats or on moist ground where the animals happened to pass. Dinosaur tracks were first discovered in the Connecticut valley, although at first it was thought they were imprints of birds because they were three-toed.

Careful study of rock formation reveals the fact that certain forms are found in certain layers and, with the help of geologists and physicists, acceptable estimates of the age of many deposits of rock containing fossils can be made. Often a continuous series of fossil forms can be pieced together, as in the horse or elephant, giving details of how the animal changed and developed during the changing periods of the Earth's history.

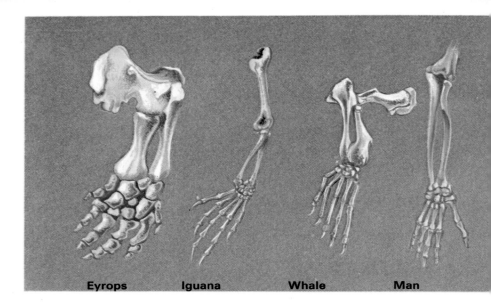

| Eyrops | Iguana | Whale | Man |

Comparative anatomy and vestigial organs

Comparative anatomy reveals similarities in the basic structures of many organisms. These similarities are of two types.
They are either analogous or homologous structures. In different animals analogous structures serve the same purpose but
are not related structurally and developmentally. For example,
the wings of insects, birds, bats and pterodactyls are analogous
structures. The wings of all these forms are used for flight, but
they have a different developmental history and internal structure. Homologous structures are related historically and developmentally and may or may not be used for the same purpose.
The front limbs of whales, men and birds are homologous; they
are all related in internal structure and in development, but they
perform the different functions of swimming, grasping and
flying.

Structural plans based on a homologous relationship nowadays are the basis of the classification of living things. On
studying the variety of animal and plant life one finds that each
group is characterized by a common form which underlies the
group's structure.

Vestigial structures often appear to be rudimentary or
underdeveloped and of no value to the owner. On studying
related forms they are clearly homologous with structures
most often of great use in other organisms. Snakes, of course,

24

Pteradactyl   Bird   Bat

Homologous variation in pentadactyl limbs (*opposite*) and analogous similarities in wing form (*above*)

have no visible external signs of legs, but internal remnants
of hip-girdles and hind limbs indicate resemblances to the
four-legged reptiles.

Similarly, the whale exhibits well-developed forelimbs in
the form of flippers, which can be related to the pentadactyl
limb and forelimb of land mammals. However, there are no
hind limbs visible, but on dissection a few bones are found
embedded in flesh, sometimes in the form of a vestigial pelvic
girdle with thigh bones attached, indicating that the whale
evolved from terrestrial mammals.

Vestigial limbs
of a python

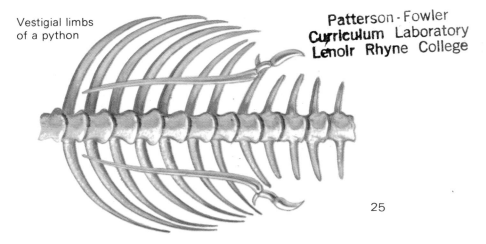

Patterson - Fowler
Curriculum Laboratory
Lenoir Rhyne College

25

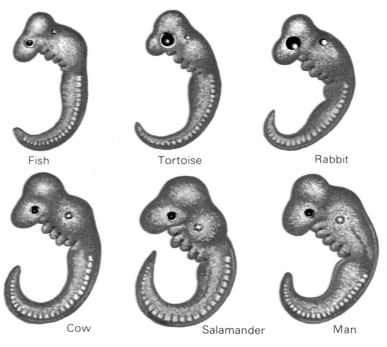

| Fish | Tortoise | Rabbit |

| Cow | Salamander | Man |

Similar embryos of widely differing adult forms

Embryology

Embryology reveals remarkable similarity of structure between animals in which the adult stages are as different as fish, tortoise, hen, rabbit and man. No one mistakes the adult forms, but it is very difficult to identify embryos without knowing the parents. Von Baer was the first to record observations of similarities, but Darwin was the first to point out the reason. If embryos are similar they must have descended from a common ancestor, from which they have inherited the embryonic stages repeated in their own development. It was suggested that during its embryonic development every individual goes through the stages by which it evolved from a primitive form. Today this is considered an exaggeration but in the embryos of most vertebrates, at comparable stages, they all have fish-like characteristics, with gill pouches or slits in the head. Each develops a vascular system with a single circulation, and the heart is not divided into right and left halves. The gill pouches are more aptly called visceral clefts as they never truly function as gills but are served in the early stages by arteries undoubtedly homologous with those of an adult fish. As each embryo completes its developmental period, each emerges as an unmistakable adult form.

Intermediates

Comparative anatomy studies have helped to form plant and animal life into related groups, and in some cases evidence has been found of certain forms which link one group to another. Some animals cannot be placed definitely in one group as they show characteristics of two groups. They are thus termed intermediates.

The arthropods and the worms can be linked in this manner by the caterpillar-like *Peripatus*. From external appearances it looks half-way between a grub and a worm. The worm-like body bears flexible appendages, but they are not jointed as in the arthropods. Internally it has the arthropod feature of a body cavity containing blood, but it is worm-like in that it has a series of nephridia, similar to the nephridia found in each segment in worms.

Peripatus

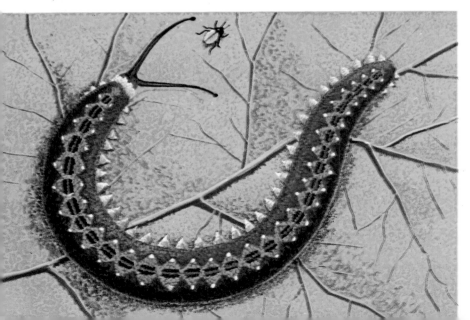

Domestic varieties of dog artificially bred by man all belong to one species.

Immunological evidence

Comparison of various animals can be made biochemically as well as morphologically. One way of studying the degree of relationship of one animal to another is through comparative immunization. This is done by injecting a sample of the blood of one animal into another animal. The animal receiving the injection acquires the ability to precipitate the proteins of the type injected in the same way that we develop immunity to an infectious disease such as measles after receiving an innoculation for the disease. The blood of the injected animal will then also precipitate proteins from the blood of the animal from which it originally received the injection and other animals related to it. The degree of precipitation often coincides with anatomical relationships.

Domestic animals and plants

Domestic dogs, cats, horses, cattle and fowl dramatically illustrate the point that animals can change their form and structure. The collie, greyhound, bulldog, dachshund and poodle would certainly be classed as separate species if found in the wild. However, they freely interbreed if allowed, and as mongrels show all characteristics intermediate between the different breeds, they are actually all one species.

Domestic plants have similarly been bred for their particular characteristics; the numerous varieties of roses are an example of this. Also, strains of wheat have been developed for their grain yield and rust resistance.

Man has, of course, selected characteristics that are of use to himself. Nature has selected characters that are useful to the organisms themselves. Therefore, while the two processes are the same in principle, they differ in direction. Because of the selection by man, however, many domesticated animals have lost the characteristics which would allow them to survive in the wild without the help and protection of man.

Classification

Systematics is the study of animal and plant classification. If all species were originally created once and for all distinct, there should be no difficulty in making lists of them. This is not so, however, as species show variation within themselves and if they spread into new environments, they may become so well adapted that they eventually form different species. Examples should be visible, therefore, of evolution in progress now, especially for those species with a wide geographical range. It is difficult to decide between variety and species, however. The cat family Felidae, for example, once comprised 25 genera, whereas today it contains only one or two, depending on the authority.

Normal variation within the cat family Felidae

Lion

Tiger

Leopard

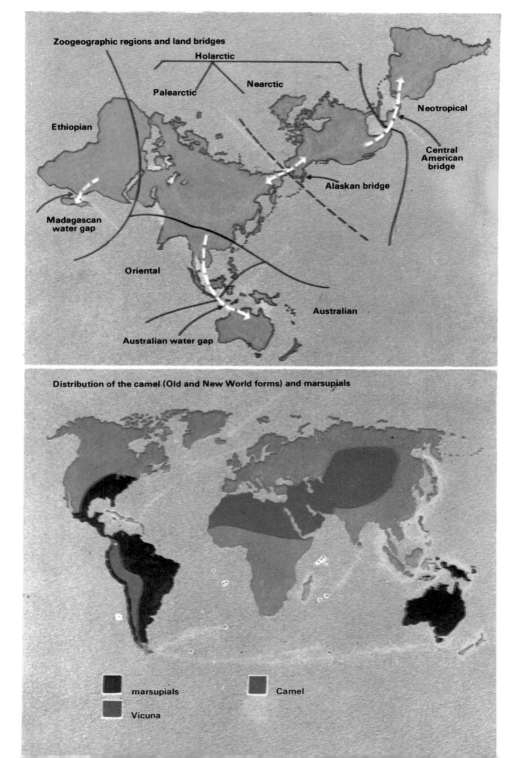

Zoogeographic regions and land bridges

Holarctic

Palearctic

Nearctic

Ethiopian

Neotropical

Central American bridge

Madagascan water gap

Alaskan bridge

Oriental

Australian

Australian water gap

Distribution of the camel (Old and New World forms) and marsupials

marsupials

Vicuna

Camel

Geographical distribution

Evidence of great importance is provided by the facts which emerge from studying the geographical distribution of plants and animals. Regions differ, sometimes very markedly, in their fauna and flora, even though they have a very similar climate. Darwin pieced together information concerning the geographical distribution of various fossil and living animals. He noticed how some related animals were separated by thousands of miles or by a mountain range or some other physical feature. The discontinued distribution of related forms must have risen when animals migrated to their present habitats and became extinct during time in the intervening regions. Fossil evidence has helped to support this view as many fossils linking the groups have been discovered in different parts of the world.

Cosmopolitan groups (those spread all over the world with no barriers to check their movement) are small in number. For example, because certain insects, bats and birds can fly, they have a widespread distribution. However, most species today have a definite area of distribution, due to the formation of barriers set up by the Earth's surface and climate.

Although the main land masses varied little during the Tertiary period, there were considerable changes in communication between them, both by the making and breaking of narrow land bridges and by sharp temperature gradients and desert areas. The various zoogeographical regions and the land bridges that allowed animal migration are marked on the map on the opposite page.

The Ethiopian region was separated from the Palearctic region by the change in temperature and desert conditions in North Africa. The isthmus between North and South America was broken in the Eocene period until the Pleistocene. Thus, the Neotropical fauna show certain distinctions from those of other regions. Australia has been cut off from the rest of the world since the late Cretaceous.

There is evidence that the many forms of animal life evolved in the central Holarctic areas and migrated away toward the extremities. Several of the types which evolved quite early remain as vestiges at the outer areas of the southern land masses, such as the marsupials, monotremes and lungfish. The distribution of the camel, vicuna and marsupials is shown.

MECHANISM OF EVOLUTION

Gregor Mendel

The name 'Mendel' is usually associated with 'genetics'. Who was this man and how did he fit into the evolutionary scene? Gregor Mendel made his discoveries around 1860, although his work remained unnoticed until the turn of the century, nearly 20 years after his death.

Johann Mendel was born in 1822 in the Silesian village of Heizendorf where his parents were peasant farmers. His early schooling and university training showed him to be a brilliant young man, and he became devoted to his philosophy studies at Olmütz University. Due to lack of food because of money shortage and overwork Mendel suffered two breakdowns in health. It became agreed in 1843 between Johann Mendel, his family and his Professor of Physics, Friedrick Franz, that he should become a novice in the monastery in Olmütz. On October 9th, 1843, he was admitted under the name of 'Gregor'.

Life was still difficult and Gregor Mendel had many physical and mental troubles during his monastic life. He became a lay brother and then a fully qualified cleric in the town of Brünn. He also distinguished himself teaching in the local school. From October 1851 to August 1853 under his professor's direction he once again became a student, this time at Vienna

DeVries, with others, later acclaimed Mendel's work.

Mendel (*opposite and below*) From the results of his plant breeding experiments Mendel proposed two laws of genetics

University, with the purpose of obtaining a more thorough grounding in the discipline of natural science.

He returned to teaching but also began to experiment with plant breeding. From 34 varieties of edible peas he selected 22 for experiments to determine the statistical relations of the various hybrid offspring. Mendel's whole theory simply predicted the number of different forms that would result from the random fertilization of two kinds of egg cells by two kinds of pollen grains.

Having satisfactorily concluded his experiments, Mendel put his findings into two papers entitled 'Experiments on Plant Hybridization'. No stirs were caused when he read these to the Brünn Natural Science Society in 1856. When he was appointed Abbot of the Monastery in 1868 he found little leisure time to continue his research, and in January 1884, he died.

His work lay neglected until 1900 when three independent investigators, DeVries in Holland, Correus in Germany, and Tscherenmak in Austria, found Mendel's forgotten paper and proclaimed its importance.

Mendel's first law

Mendel set about investigating the problem of the manner in which free-breeding variations within a species are related to one another. In a number of garden plants he noticed that within each species certain definite variations could be observed. In the garden pea, *Pisum sativum,* he saw many characteristics: for example, there were tall (6 to 7 feet) and dwarf (9 to 18 inches) varieties; their seeds could be smooth and rounded or wrinkled in appearance; the unripe pods could be green or yellow. His approach of basic biometric statistics fitted in with his physics background and the results had to be analyzed and mathematically interpreted. Mendel knew the life histories of the plants and thus in one experiment crossed a dwarf variety about a foot high with a tall variety reaching about 6 feet. He obtained hybrids that were all tall. He pollinated the hybrids with their own pollen and obtained 75 percent tall and 25 percent dwarf, that is a ratio of three to one. When he self-pollinated this second generation he found all the dwarf variety bred true in the following generations. However only one-third of the tall variety bred true, the rest gave 75 percent tall and 25 percent dwarf again, and so the process continued.

From many experiments using different characters Mendel realized there must be some characters (such as wrinkled seeds or dwarfness) which can, during mating, be 'masked' or overcome when in the company of certain other characters (such as round seeds or tallness). Although 'masked' in some given generation, they retain the power to reappear in a later generation. For these characters he used the term 'recessive'. Those characters that could mask the recessive characters he called 'dominant'. If the two were coupled together the dominant character only would be visible and has the effect of preventing the recessive displaying its effect or form.

Thus, from the third generation results Mendel worked out that one-third of the tall plants had no dwarf factors being masked and therefore bred true. The others would be masking the dwarf recessive factor which when the plant was self-fertilized would give the three to one ratio again.

The seed counts and interpretation enabled Mendel to propose what is known as his first law which states that *characters*

P₁

F₁ generation

F₂ generation

In this cross, a character for seed form and color is suppressed in the F₂. When plants of this generation are self-pollinated, however, the character reappears in a quarter of the next generation.

are controlled by pairs of factors which do not blend during life and which pass into separate cells during reproductive processes, prior to fertilization.

Mendel's second law

Mendel did not stop after investigating just one pair of contrasting characters; he went further, noting the results when plants with two pairs of factors were self-fertilized.

One must remember that although Mendel was not familiar with chromosomes, he saw the need for finding out how characters would behave in relation to each other in their passage from generation to generation.

In one study, he chose seed form and seed color as his subjects, using a pure breeding plant with round and yellow seeds as one parent, and a pure breeding plant with wrinkled and green seeds as the other. His method was the same as that previously described, but he recorded the frequencies with which each character appeared with the other, instead of just separate frequencies. His results showed that the first filial generation from all the plants bred were round and yellow seeds, indicating the dominance of round over wrinkled, and yellow over green characters, respectively. After self-fertilization of these seeded plants, Mendel found he had a second filial generation showing four different combined sets of characters.

He calculated the proportions and found them approximately in the ratio of nine round and yellow seeded plants, three round and green seeded plants, three wrinkled and yellow seeded plants and one wrinkled and green seeded plant.

From these results, Mendel deduced a second law of inheritance. He stated that *each of a pair of contrasted characters may be combined with either of another pair during the process of reproduction so that there is complete independence of combination among the factors present.*

In all, Mendel studied seven pairs of characters in peas, involving seed color, seed surface, flower color, vine height, color of unripe pods, pod shape and position of flowers. These crosses which involved two character differences separable in inheritance are called di-hybrid crosses (a cross involving a single pair of alleles is monohybrid). Similar experiments of di-hybrid inheritance have been carried out in *Drosophila,* the fruit fly, with the same ratio results. The particular combination in which characters are brought into a cross makes no difference at all to the manner in which they

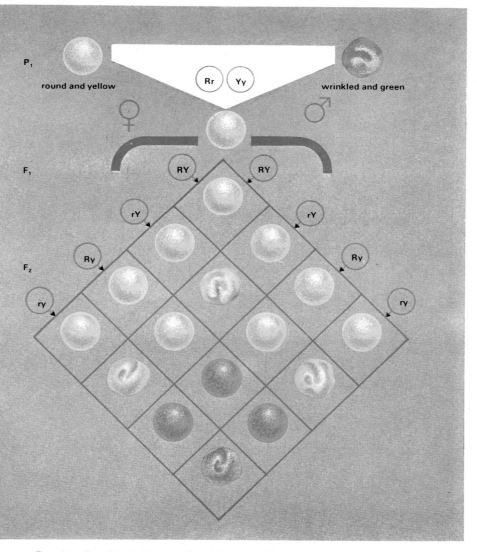

Results of a di-hybrid cross involving seed form and color.

are assorted and recombined in the F_2. If both dominant characters are brought in by one parent and both recessives by the other, the ratio results are the same as when both parents have one dominant and one recessive. The results are always 9/16 dominant characters, 3/16 one dominant one recessive, 3/16 other dominant other recessive, 1/16 recessive.

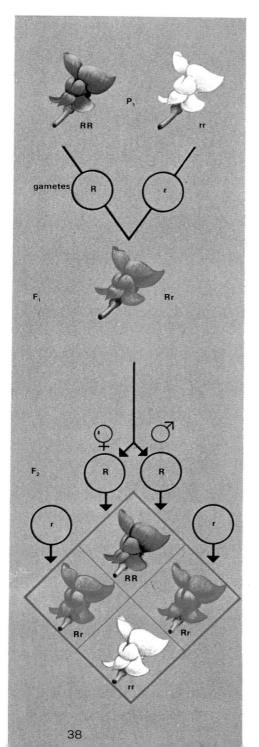

Incomplete dominance

Not all characters behave as perfectly as Mendel's laws suppose.

Incomplete dominance indicates an apparent failure of one allelomorphic character to dominate the other in the F_1 generation. In the Andalusian domestic fowl, it is possible to obtain true breeding strains of the black and white forms. When crossed, the first generation offspring are neither black nor white, but an intermediate color called 'blue' or 'splashed white'. If these fowls are allowed to breed among themselves, the next F_2 generation shows a segregation into blacks, blues and whites in the proportions 1:2:1. It is now known that if a gene for black color is present with a gene for white color, then a single black gene is unable to produce full black pigment development. A fowl must possess genes from both parents for full black pigment production.

In plants, the cross between red and white snapdragon (*Antirrhinum*) gives hybrids of an intermediate pink color. This again shows that when the alleles R and r come together in a cross neither is dominant, and the heterozygote Rr is pink flowered.

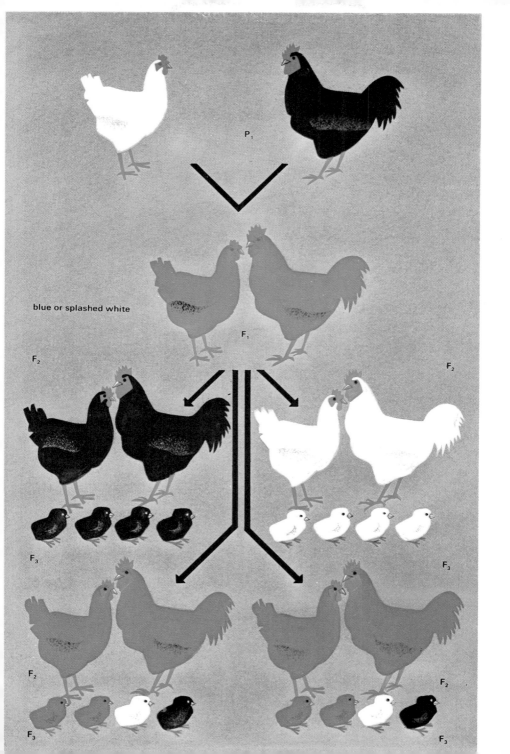

P₁

blue or splashed white

F₁

F₂

F₂

F₃

F₃

F₂

F₂

F₃

F₃

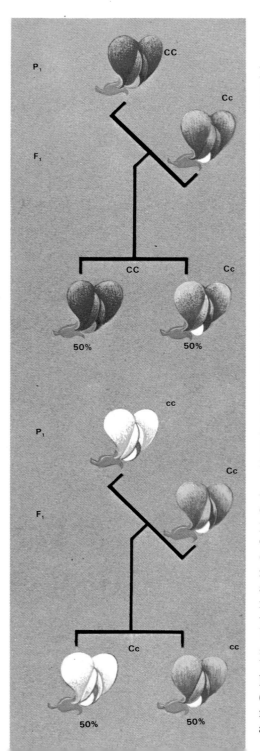

Back crosses

Mendel also examined allele behavior in back crosses. A heterozygous F_1 individual is crossed back to either of the homozygous parents. The red-flowered F_1 hybrid from the cross of red and white peas is crossed back with the dominant red-flowered parental variety in one, and the recessive white-flowered parental variety in the other. The red-flowered F_1 hybrid, according to Mendel, is a heterozygote Cc and the white-flowered parent a homozygote cc. The gametes produced by a hybrid heterozygote are always pure. Therefore, in a red-flowered F_1 plant one half should carry C and the other c. The gametes of a white-flowered plant which are recessive are all c.

The progeny in the back cross to the red-flowered parent are all red-flowered plants and if Mendel's theory is correct they should consist of 50 percent homozygous red-flowered (CC) and 50 percent heterozygous red-flowered (Cc). This is verified by looking at the back cross to the recessive white-flowered parent. White-flowered (cc) and heterozygous red-flowered (Cc) are produced in equal numbers (see illustration at left).

Lethal crosses

According to Mendelian segregation, gametes that carry different alleles of the same gene unite at random and give a ratio of three dominants to one recessive. However, these conditions are not always realized.

Matings between mice with yellowish fur have produced offspring with yellow and non-yellow fur in a ratio of two to one. Furthermore, matings between yellows give litters which are smaller in number by about one quarter than litters from yellow and non-yellow parents. Apparently zygotes homozygous for yellow are inviable, and this hypothesis has been verified by scientists crossing yellow females with yellow males. They found that some embryos die in an early stage of development. From the illustration:

$$Ay = \text{a dominant gene for yellow fur}$$
$$a = \text{its recessive allele for non-yellow}$$

The allele Ay is lethal when homozygous but when present with the non-yellow recessive allele a, it merely modifies the color of the fur to a yellowish tinge.

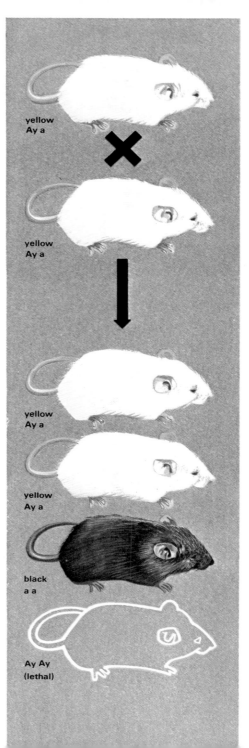

yellow
Ay a

yellow
Ay a

yellow
Ay a

yellow
Ay a

black
a a

Ay Ay
(lethal)

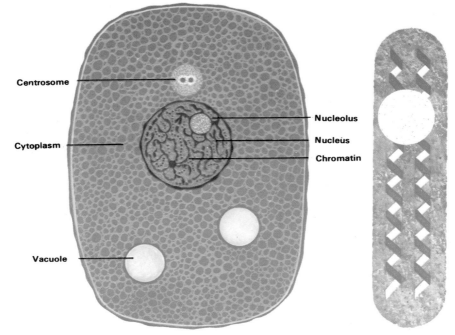

Centrosome

Cytoplasm

Vacuole

Nucleolus

Nucleus

Chromatin

Diagrammatic cell and chromosome

Cells and chromosomes

The living substance of both plants and animals is organized
into microscopic units called cells. Cells were first seen by
Robert Hooke as early as 1665, when he observed and described the cells of a piece of cork.

Usually the cell consists of a dense body, the nucleus, en-
closed in its own membrane and surrounded by less dense
colloidal fluid, the *cytoplasm*. The cytoplasm contains a
variety of structures including the *mitochondia,* centers of
enzyme activity and production, only just visible through a
light microscope. Animal cells contain *golgi bodies,* whose
function is uncertain and *centrosomes,* important in cell division. In the nucleus are found the structures of great genetic importance, the *chromosomes*.

Cells of the body carry out division for either growth by
mitosis, or for the formation of sex cells by *meiosis.* Mitosis
is the process by which a cell divides into two parts during
growth, during which the chromosomes undergo definite

changes. These are the recognized stages:

Interphase. The cell is not dividing and separate chromosomes are not usually distinguishable.

Prophase. The cell is preparing to divide and the chromosomes become clearly visible as threads which gradually shorten and thicken by coiling.

Metaphase. The nuclear membrane disappears and the spindle appears. Arranged around the central plane of the cell, each chromosome splits longitudinally. Each cell now has, in effect, a set of 'double' chromosomes.

Anaphase. The centromeres split and the chromatids move toward the poles.

Telophase. The chromosome halves move to the poles forming one complete set at each end.

The cell now finally divides, the spindle disappears and the chromosomes lengthen. 'Daughter' nuclei form, each containing a new set of chromosomes and the conditions of interphase are restored.

Stages of mitosis illustrated diagrammatically

prophase

metaphase

metaphase

anaphase

telophase

interphase

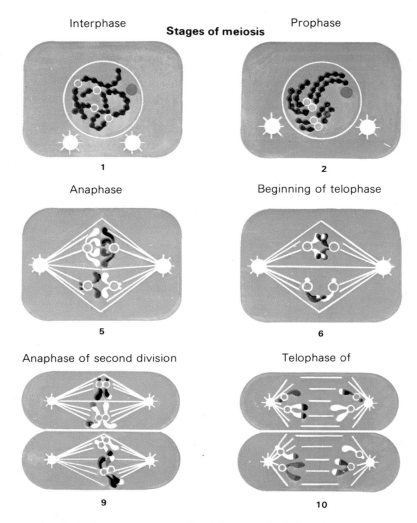

Stages of meiosis

Interphase

Prophase

1

2

Anaphase

Beginning of telophase

5

6

Anaphase of second division

Telophase of

9

10

Meiosis is not concerned with growth. The chromosome number of a species is always fixed. How is it that the union of the parent cells does not produce a doubling of the chromosome number? At some point during sexual reproduction, the chromosome number in the sex cells has to be halved and this happens during the formation of the male sex cells *spermatozoa,* and the female sex cells *ova,* by a special method of cell division called meiosis.

Meiosis differs from mitosis in that it produces four new

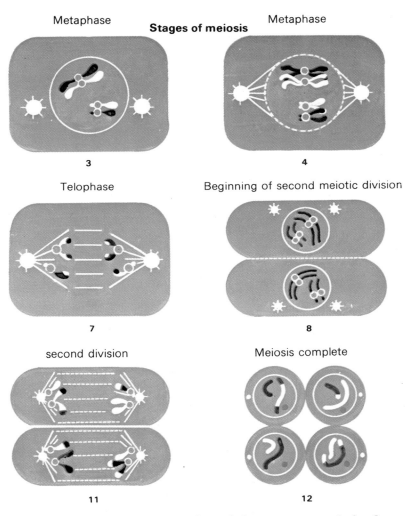

Metaphase

Stages of meiosis

Metaphase

3

4

Telophase

Beginning of second meiotic division

7

8

second division

Meiosis complete

11

12

cells in each of which the number of chromosomes is halved. The process is brought about by two nuclear divisions, but only one duplication of the chromosomes. We refer to these two divisions as the first and second meiotic divisions. During the first meiotic division, the phases occur in the same sequence as they do in mitosis. However, in the prophase of meiosis, there is a significant difference in the behavior of the chromosomes. A number of distinct stages within the first meiotic division can be recognized.

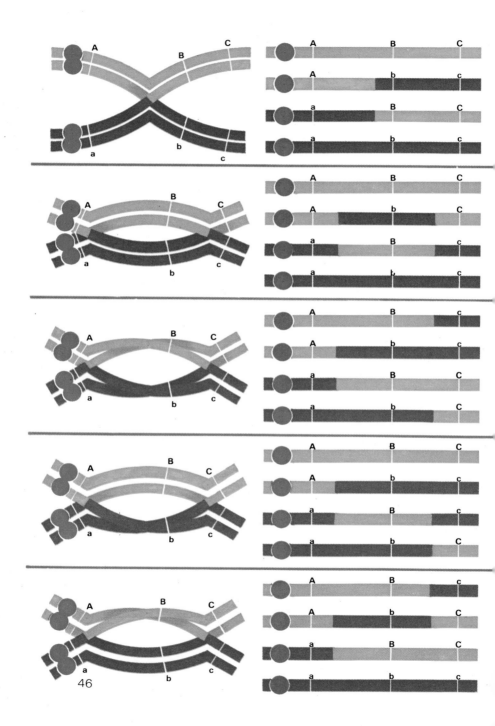

Linkage and crossing-over

It is now known that the chromosomes carry the genes responsible for the inheritable characteristics. As the fixed number of chromosomes for each species is quite small, each chromosome must carry a fantastic number of genes. The genes are arranged in linear order, along the length of the chromosomes, and in 1910 Morgan discovered many facts about their positions. He believed that the tendency for linked genes to remain in their original combinations was due to their residence on the same chromosomes. He also thought that the degree or strength of linkage depends upon the distance between the linked genes in the chromosome. This basic idea has now led to the construction of genetic or linkage maps of chromosomes showing the exact position on a chromosome in which an inheritable characteristic is found.

If the genes are always in the same place, then two genes situated on the same chromosome should remain together in all cases. However, although genes are bound together, the chromosomes when pairing during meiosis can exchange genes by 'crossing-over'. This interchange of sections of chromosomes leads to the recombination of linked genes and so some alleles switch from one chromosome to another. The chromosomes pair with remarkable preciseness during the prophase of meiosis, and this is evidently brought about by mutual attraction of the parts of the chromosomes that are similar because they contain allelic genes. Between the *pachytene* and *diplotene* stages the paired chromosomes each divide into two chromatids so that there are four in number.

At the first division of the chromosomes, the newly formed chromatids exchange one or more links or *chiasmata*. At each chiasma, two of the four chromatids become broken and then rejoined so that the new chromatids are compounded from sections of the original ones. The new chromosomes that arise as a result of meiosis carry genes that before meiosis were located in different members of the pair of chromosomes. However, crossing-over does not occur in the majority of gametes. Usually, about 97 percent of the gametes contain the parental combination, and 3 percent contain recombinations.

Chromosome cross-overs and resulting gene recombinations

Determination of sex

How is it every child has a mother and father, but some children are males and others are females? What determines whether an offspring will be a male or a female? Questions like these puzzled non-biologists and biologists alike for centuries, and before 1900 most theories were just wild guesses.

Heredity was thought to determine merely the similarities between parents and offspring, and the heredity of a child was supposed to be a compromise between the inheritable characteristics of the parents. How did sex fit into such ideas of heredity? Could there by any relation or control between the two? The vital evidence necessary to answer the sex determination question was not unfolded until the early 1900's by genetic researchers.

It was observed in some insects that male chromosomes gave an odd total number, and that females had an even number. Another observation was that in some males there was often one pair of chromosomes made up of unequal part-

Unequal chromosome pair (*left*) and normal pair (*right*)

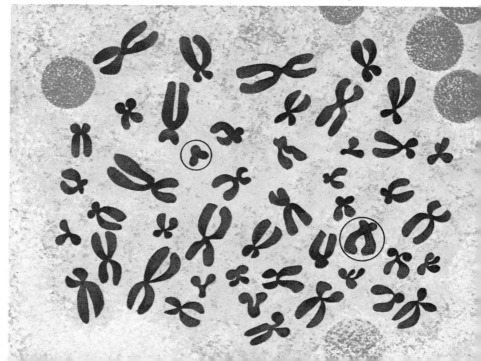

ners, whereas in the female this pair was formed by equal chromosomes. The unequal chromosomes are called the *XY* pair and the matching chromosomes the *XX* pair. When meiosis is completed in the male, there are two kinds of spermatozoa, one with *X* chromosomes, the other with *Y* chromosomes. In the female only a single egg with an *X* chromosome is present. The sex of an individual is, accordingly, determined at fertilization. If an *X*-carrying spermatozoon fertilizes the female egg, the union will produce a *zygote* with two *X* chromosomes, developing into a female. A *Y*-carrying spermatozoon gives an *XY* zygote which becomes a male. The chromosomes obviously differ in the male and female and so it follows that the genes will differ.

In man, other mammals and most insects, the heterozygous sex is the male. However, although the principle is much the same, the chromosome mechanism of sex determination has been found to vary in different organisms. In birds, butterflies, moths and some reptiles and amphibians, the *XY* zygote produces a female and an *XX* zygote produces a male.

Mechanism of sex determination in human offspring

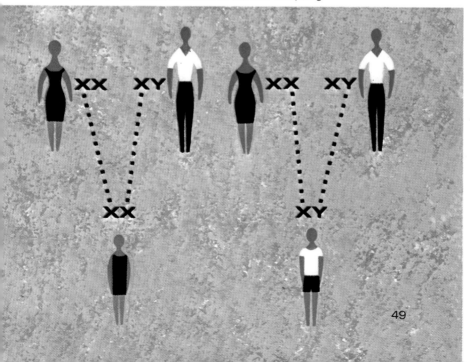

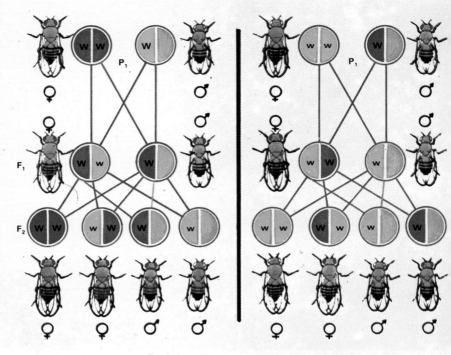

Reciprocal crosses showing sex-linked inheritance of eye color in *Drosophila*.

Sex linkage

The sex chromosomes, it was found, did not just contain genes which controlled the sex of the individual. They had many genes concerned with other characters of the individual. Morgan was the first to discover this from his work with the fruit fly, *Drosophila*.

Morgan found one individual, which had white eyes, breeding with normal red-eyed wild *Drosophila*. The white-eyed *Drosophila* bred true in subsequent generations, and a new variety was established. He crossed this new white-eyed variety with the wild red-eyed type. The results from a cross of a white male and a red female were different from those obtained from the reciprocal cross of a red male with a white female.

From the cross of white-eyed male with red-eyed female, the first generation all have red eyes. When these are bred together, one-quarter of the offspring have white eyes, indicating that red-eyed allele is dominant over white-eyed. But of the F_2 offspring only half of the males are white-eyed and the others are red-eyed, but all the females are red-eyed. Although

50

the females do not show any external differences, their sex chromosomes contain two different kinds of genes. This is shown when the two types of females are bred as all the offspring are red-eyed, except for about one-eighth of the total which are white-eyed males. This results from the half of the females carrying the recessive white-eyed gene which appears when paired with a male sex chromosome.

Different results can be expected when a red-eyed male is bred to a white-eyed female. By studying the illustration on the opposite page it can be seen that the F_1 offspring consists of the red-eyed females and white-eyed males. These produce F_2 offspring of red-eyed and white-eyed flies in equal numbers in both sexes. All the white-eyed flies breed true, as do the red-eyed males when bred to pure red-eyed females. But red-eyed F_1 females must be heterozygous for when bred to either white-eyed or red-eyed males, half the male offspring are white-eyed.

Studying the *Drosophila* crosses it can be seen that the sex-linked trait of white eye color follows a criss-cross inheritance. A male transmits his sex-linked trait to only his daughters where it is recessive.

DNA—mechanism of inheritance

Darwin had proposed his theory of evolution without taking into account the mechanisms of inheritance. This was a vital link in the theory because if the characteristics of the new generations were always a mixture of those of their parents, natural selection would tend to reduce variety rather than increase it. In an empirical way, Mendel described the way in which characteristics or traits are handed down. Mendel showed that a character was represented by some structure in the organism and was preserved without being blended with other characters. This is not to say that many characters did not appear to be blended, as with the pink sweet pea described earlier (see page 38). In this case and in others in later generations, the original character showed up unaltered. This particular nature of inheritance was a necessary condition for Darwin's theory to work.

It was many years later, after the theory of evolution was finally accepted by most biologists, that the mechanism for inheritance was proposed. Francis Crick and James Watson were the scientists who first described the structure of the material governing inheritance, or DNA.

The basic molecular structure of hereditary material is a subject of its own, and only a brief summary of the mechanisms of inheritance will be given here.

The deoxyribonucleic acid, or DNA, ribbon contains thousands of units. These units are called nucleotides. A nucleotide is made up of one of a number of bases attached to a sugar molecule—deoxyribose—and a phosphate molecule. The deoxyribose and the phosphate molecule form the sides of a ladder-like arrangement of the DNA while the cross-pieces are made up of paired bases. This whole structure is normally twisted into a spiral (see illustrations on opposite page).

DNA is found in the nucleus of the cell. During cell division the DNA spiral unwinds and splits into two pieces. Each side then pairs with new molecules, thereby forming two sets of molecules, both having the same sequence of nucleotides. As the cell divides, one set goes with each of the two new cells.

There are four different kinds of bases: two purines, adenine and guanine; and two pyrimidines, thymine and cytosine. A purine is paired with a pyrimidine, usually adenine with thymine and guanine with cytosine. The bases pair in the following four combinations: A-T, T-A, G-C, and C-G. These pairs form a code, and three of the pairs together determine the position of amino acids in a protein molecule.

Proteins are made up of 20 amino acids and the kind of protein is determined by the kind and order of amino acids in the sequence. Any change in the nucleotide sequence in DNA will produce a change in the amino acid sequence in a protein, so that a different protein will be produced.

Proteins are the basic building blocks of organisms. In the form of enzymes they control chemical reactions. In the form of structural proteins they form most of the important structures of an organism.

Mutations are a result of a change in the hereditary material, DNA. Most commonly, mutations are caused by a substitution of one of the nucleotide pairs, thereby bringing about a change in the resulting protein. Mutations are the ultimate source of genetic variability.

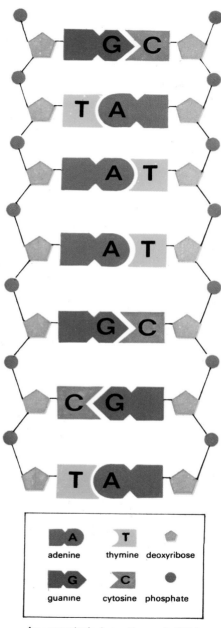

An unspiraled section of DNA showing the many nucleotides, each made up of a base, a deoxyribose, and a phosphate.

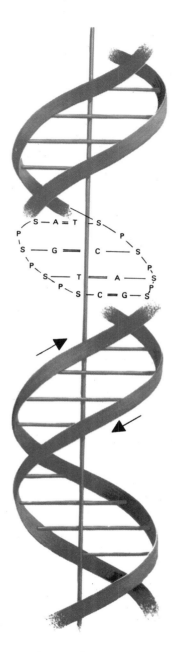

Diagram of a small section of a DNA molecule.

Altitude-induced variation in *Horkelia california*

Variation

Although mutations supply the basic source of variation in a population, the variety is greatly increased by genetic recombination—that is, various sets of different characters are combined in different ways in different organisms. These recombinations in the genes are brought about through crossing over during meiosis (see pages 46–47).

Variation may also be caused by changes in the environment. These changes, or modifications, are not inherited, however, and play no part in evolution.

Transplanting experiments often illustrate the effect of the environment on a plant's growth. A single plant is divided into pieces so that a number of plants of the same age and genetic constitution are obtained, forming a *clone*. The individuals of the clone may be grown in different climatic conditions in order to study the effects of the environment.

Transplanting experiments of *Horkelia californica* to different altitudes produce interesting results. This plant, which grows wild at 600 feet in California, reaches a good height at sea level, but is remarkably stunted in its growth at higher altitudes, often perishing during the winter months.

The amount of food an animal or plant consumes or builds up is another extremely important environmental factor which helps to produce variation. Pigs are an example. Those able to eat as much as possible get very large and fat, while those put on a diet are long and lean. This experiment

54

was performed in a pure strain of pigs where hereditary variation can be discounted. Some were given restricted amounts and others unrestricted amounts of the same food for equal periods of time. Differences were found in the development of bones, those on the unrestricted diet at 16 weeks being much larger with massive bones. When full grown, it was discovered that those on a restricted diet had longer and narrower skulls. They also had a longer lumbar region of the vertebral column (loin) and larger hip girdles than the unrestricted pigs, although it took them longer to become fully developed.

Genetic isolation

If certain groups of a population become isolated by geographical, climatic, or other factors so that the groups can no longer breed with each other, the forces of natural selection may select one or another of the variations in the isolated population so that they diverge sufficiently from the original group to form a new species.

Pigs fed on unrestricted diet (*top*) and restricted diet (*bottom*)

Geographical isolation

The birds of the Galapagos Islands provide an excellent illustration of the role of geographical isolation in the formation of new species. The Galapagos are made up of 13 larger islands separated from each other by distances up to 100 miles. These volcanic islands are relatively new, geologically speaking. The animals and the plants found on them reached the islands from the coast of South America by chance. Because of the great distance from the mainland (over 600 miles), only seven species of land birds reached the islands. One of these species was a finch. Today, the descendants of this one species of finch are represented by 14 different species. The main differences are found in the form of their beaks, which varies with their food habits.

The Galapagos ground finches closely resemble the South American finch species which feeds mainly on seeds. Two of these have left the ground to feed on the prickly pear cactus, *Opuntia*. Those living in trees have beaks adapted for insects, although one is a vegetarian. Among this group is a woodpecker-like finch, *Camarhynchus pallidus*, which climbs trees woodpecker-fashion and digs insects from the bark with a stick. The warbler finch group, Certhidea, have long slender beaks for eating small soft insects and have behavior habits similar to the true warbler. They have evolved to fill this role because until recently there had been no warbler species on the islands. The fourth group has only one member, the Cocus finch, found only on Cocus Island.

Populations of the early species of finches varied slightly on the different islands and were changed by natural selection. Occasionally, by chance, a pair of birds from one island would cross to a neighboring island. If the new arrivals were already sufficiently changed so that they were not recognized as the same species, then the two groups would not interbreed and the island would have two separate species, each of which would still continue to change in its own direction.

The groups of finches (*top*) observed by Darwin in the Galapagos Islands (*opposite*)

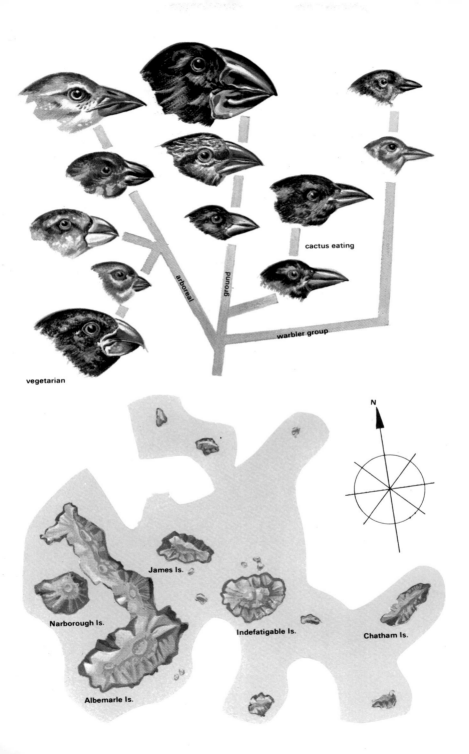

cactus eating

arboreal

ground

warbler group

vegetarian

N

James Is.

Narborough Is.

Indefatigable Is.

Chatham Is.

Albemarle Is.

Geographical variation

When a species covers a large area, on close observation it is found that it gradually changes in certain characters over different parts of the range. The change may be in the size, color, shape or other features. Such gradients of characters are called *clines,* a term introduced by Sir Julian Huxley, a well-known contemporary biologist and evolutionist.

The lesser black-backed gull and the herring gull occupy a ring-shaped range around the North Pole. On the shores of the Arctic Ocean in Eastern Siberia lives the Vega gull (*Larus argentatus vegae*) with dull flesh-colored legs and a dark mantle. Eastward across the Bering straits to North America, the Vega gull grades into the American gull (*L. a. smithsonianus*) and across the Atlantic into the British herring gull (*L. a. argentatus*) with flesh-colored legs and a light mantle.

In a westerly direction from Siberia, the Vega gull grades into the lesser black-backed gull (*Larus fuscus fuscus*) and

Gulls showing graded characters over their range

Vega

Lesser Black-backed

British Herring

Scandinavian Black-backed

American Herring

Size cline of the puffin

its British race (*L. f. graellsii*). Thus, there is a continuous gene-pool of gulls around the North Pole, with the gulls inter-breeding around the chain. However, the ends overlap in Britain with the lesser black-backed gull and the herring gull living together, but differing in behavior and appearance. The lesser black-backed gull breeds inland and is migratory in winter; the herring gull nests on cliffs and is resident. They do not interbreed, behaving as distinct species.

The puffin provides a good example of a size cline. From the Balearic Islands in the Mediterranean to Spitzbergen in the Arctic, the puffin varies in a gradual change. The maximum wingspan of these birds increases at the rate of 1 percent of linear measurement per 2° latitude north.

This is also an example of Bergman's Rule, according to which warmblooded animals increase their size in colder climates. This is an adaptation to reduce heat-loss, since the larger the animal, the smaller is the ratio between its volume (which produces heat) and its surface (which loses heat).

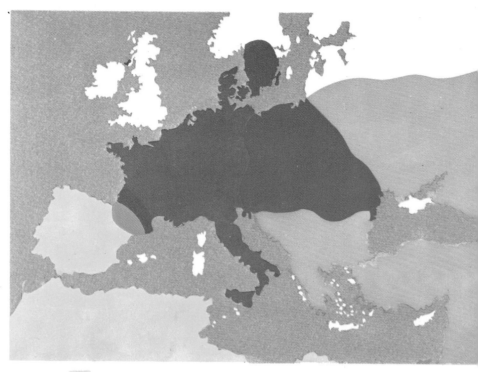

} *Rana ridibunda* (western)
} *Rana ridibunda* (eastern)

} *Rana esculenta*

Ecological isolation

Biological isolation is a necessary condition if species are to diverge and new species form. As we have seen, it may be geographical when physical features such as mountain ranges separate whole populations. Also extremely important is ecological isolation, where different modes of life such as feeding, breeding, and nesting habits, or adaptations to different temperatures or degrees of salinity, separate portions of populations.

Rana esculenta and *Rana ridibunda*, species of edible frogs, provide an example of overlapping geographical distribution. The range of the latter overlaps the former on both sides with an Eastern race and a Western race. The species do not normally interbreed, as there is a reproductive isolation barrier — *Rana ridibunda* breeds earlier than *Rana esculenta*.

Genetic isolation

Related plants occuring side by side may be isolated by their inability to breed and produce fertile offsping. Incompatibility between the gene and chromosome mechanisms of the two plants is often the cause of this.

When *Primula verticullata* is crossed with *Primula floribunda*, hybrid offspring are produced, but they are sterile because the chromosomes of one parent species are incompatible with those of the other. The matching of the male set of chromosomes with the female set is difficult in these conditions, and the formation of the sex cells is thrown out of gear. Occasionally, however, the hybrid plant undergoes what is known as *polyploidy*. This is a doubling of the chromosomes, and when it occurs the hybrids are able to breed with other similarly formed hybrids, but are sterile with both parent species. The hybrid then breeds true, with a different structure and habit from each of the parent species. It thus fulfills all the criteria of a new species and is called *Primula kewensis*. Similarly the peony can be found in the diploid, triploid and tetraploid form. Polyploidy may be an important evolutionary mechanism, but it occurs only in plants.

Overlapping ranges of edible frogs (*opposite*) and polyploidy in the peony (*below*)

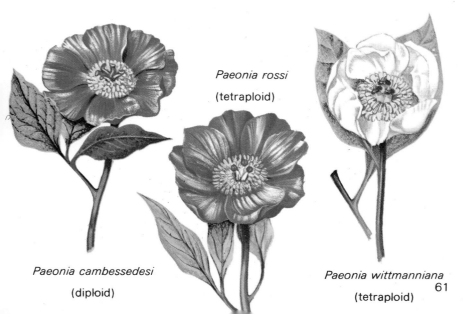

Paeonia rossi
(tetraploid)

Paeonia cambessedesi
(diploid)

Paeonia wittmanniana
(tetraploid)

61

Male lyre-bird

Sexual selection

In the breeding season of many insects, fish, birds and mammals, the males show ornamental features and displays. In only a few instances do they appear in the females. Darwin explained the varying appearances in his subsidiary theory of sexual selection. He argued that before mating occurs, the female has the choice of many suitors. She chooses one or more and each chosen will show slight differences from the rest. These characters of the fortunate chosen males will be inherited and appear in the male offspring, as the genes will be on the sex chromosomes and thus sex-linked. Darwin's argument and explanation is that these individuals with well-developed striking colors, structures or elaborate behavior benefit from greater breeding success and will therefore have more offspring. There are objections to Darwin's theory. In many cases these secondary sex characters are not the result of sexual selection alone, but also of natural selection favoring general reproductive activity through conspicuous recognition marks, virile behavior, fighting ability and so on.

This phase of selection does not work by the complete elimination of the least fit, but by eliminating reproduction of the least fit. It is most effective among the males, as the females usually manage to be fertilized, but there may be a great battle among the males for breeding territories and the possession of the females.

Two examples of sexual selection are the lyre-bird and the stickleback fish. The male lyre-bird shows his display material in remarkable specialized tail feathers. He spreads his tail forward over his head and vibrates the feathers until they resemble a shimmering haze of light and this attracts the females to him.

In spring, the male stickleback develops his courting colors of a bluish-white back and a brilliant red throat and belly. He stakes out his home territory, builds a nest, defending his rights from all other males and then tries to entice an egg-bearing female by a most elaborate zig-zag dance. If successful, the male terminates his courting behavior by fertilizing the eggs laid by the female. Therefore, genes from that individual are present in the next generation.

Male stickleback enticing female to nest

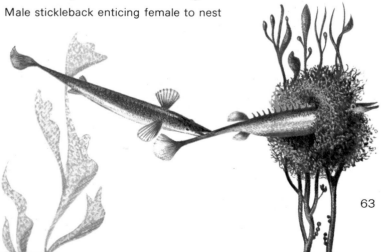

63

Adaptation

Adaptation includes those features in an animal or plant which enable it to survive in its environment.

The woodpecker, for example, has four main adaptive features. The four well-clawed toes are formed so that two face forward and two backward, so they can firmly anchor themselves to the tree; the tail feathers are stiff and can be used to help prop the bird securely against the tree; the beak is long and strong so that holes can be drilled in the bark; and the tongue is very long, enabling the bird to reach and catch grubs at the bottom of the holes.

Many other creatures have evolved adaptations for flight, for example, insects, birds and bats. In each case of adaptation the wings are of different construction, but all are successful in flight.

Many species of desert plants, although not closely related, are quite similar in general appearance. This is because they are all adapted to keep loss of water by evaporation to a minimum. They have a tough epidermis, reduced leaves, thickened stems and smooth surfaces as protection against desiccation and death.

Woodpecker (*above*) and desert cacti (*below*)

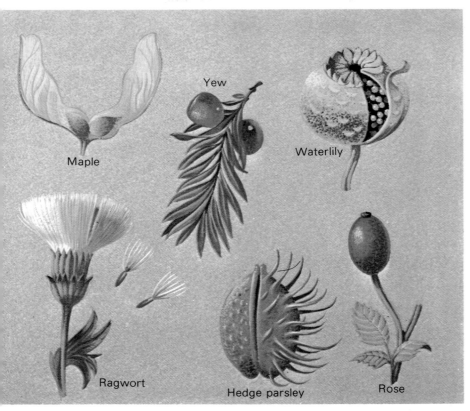

Adaptations for seed dispersal in plants

The survival of a plant species depends to a large extent on the reproductive process and the efficient dispersal of the seed. It is advantageous to the species that the seeds are carried some way from the parent plant, in order to obviate the competition for food, light and water. Many natural agents do this, such as wind, water, birds and animals.

Some seeds are so light that they are easily blown by the wind, for example those of ragwort and dandelions; other heavier seeds have structures enabling them to fly, such as the maple.

Dispersal by water is uncommon; however, in some plants — for example, the lily — the seeds have a spongy covering that enables them to float.

Many seeds have tiny hooks so that they adhere to passing animals and are rubbed off far from the parent plant. Others, such as the rose hip, are brightly colored in order to attract birds.

Heliconius pardalinus

Melinaea madeira

Melinaea maelus

Mechanitis egaensis

Ceratinia anastasia

Butterflies exhibiting Mullerian mimicry

Mullerian mimicry

Mullerian mimicry is the term applied to several different species that closely resemble each other and are all distasteful to their potential predators. For example, several species of butterflies sufficiently resemble each other so as to be indistinguishable to birds. Thus, the predator bird, after one distasteful experience, learns to avoid this color scheme, with the result that all the species benefit and the overall losses are greatly reduced. The more species in a 'mimicry ring' the greater the advantage and the less eaten, but some butterflies must be sacrificed in educating each generation of birds. The resemblance is not necessarily very exact, as it is in Batesian mimicry, as the color patterns are the reminders. A mimicry ring of South African butterflies is shown above. Mimicry is more common in the tropics than in temperate regions.

Papilio dardanus (meriones male)   Papilio dardanus (meriones female)

Hippocoon   Niobe   Cerea

Amauris niavius   Amauris echeria   Bematistes tellus

Butterflies exhibiting Batesian mimicry

Batesian mimicry

Unlike Mullerian mimicry, not all Batesian mimics are distasteful. Most species are chemically unprotected and gain protection from their resemblance to the distasteful ones. From this it follows that the mimics must never exceed a certain number relative to the models' number or the protection will be lost. This phenomenon has been built up from those variants that confer high survival value where predators learn to shun well-marked unpalatable types and natural selection pressure is exerted in favor of the closest resemblance to those types. Bates observed this initially in Africa by studying the widespread African butterfly, *Papilio dardanus*. The males of this species do not mimic and are therefore conspicuous. The females mimic many unpalatable species of models and only resemble the males in Madagascar and Abyssinia.

Light form favored on lichen-covered bark.

Natural selection

With modern knowledge in genetic, ecological, zoological and mathematical fields, the effects of an environment on a certain variant of a species can be measured. This enables a researcher to study long-term effects and to find out if evolutionary change does take place.

H. B. D. Kettlewell's research on 'industrial melanism' in moths illustrates evolutionary changes. This is the phenomenon in which moths have changed their complicated patterns from a light to an all-black coloration. The first work by Kettlewell was with the peppered moth (*Biston betularia*), but this is not the only species which has changed or is changing.

Before the Industrial Revolution the majority of trees throughout Great Britain had lichens on their barks. Now such trees occur only in unpolluted areas. Carbon dust has killed the lichens on trees in industrial areas and rendered

Melanic form favored on sooty bark.

their trunks and branches black. The typical light form of the peppered moth is well camouflaged for resting on the lichen-covered bark of trees, being almost invisible to predatory birds. Before 1848 this was the only form recorded, but soon after this a dark melanic variety, *Biston carbonaria*, appeared. Due to a genetic mutation this variety differs from *betularia* by a single recurring dominant gene slightly more vigorous than the normal light type. In the mid-Victorian days, because it was more conspicuous, the *carbonaria* variety was constantly eliminated. However, the gradual darkening of the moths' environment meant that the more conspicuous light *betularia* variety was taken by birds rather than the dark *carbonaria*.

By 1900 melanic *carbonaria* outnumbered the *betularia* form in the Manchester area. There had been a sweeping change in just 50 years. Today, only in the unpolluted areas of Northern Scotland, West Devon, Cornwall and parts of the south coast is the *betularia* variety found.

Natural selection in the land snail

The land snail (*Cepaea nemoralis*) is highly polymorphic, and the many color variations and banding on the shell have been studied in the snails' various environments by Cain and Sheppard. The shell can be yellow, pink or brown and can also be five-banded, one-banded or unbanded. All combinations of shell color and banding pattern can be found, but some are much more common than others. Brown is dominant over pink which is dominant over yellow, the three being apparently controlled by three allelomorphs. Unbanded is dominant over the presence of bands, and the locus is linked with that controlling color. The one-banded form is dominant over the five-banded form and is not controlled by the locus which determines the presence or absence of bands. The proportions of the various color and banding types differ greatly from one colony to another, and Cain and Sheppard found that they are strictly related to the ecology of the site. The most common shell coloring is that which is least conspicuous to predators against background vegetation. The most advantageous shell colors are yellow (greenish when the animal is inside) in green areas, pink on leaf litter, and reds and browns in beech woods with red leaf litter and numerous exposures of blackish soil.

Migration of snails from unsuitable to suitable areas is highly unlikely as distances would be prohibitive, and so it appears that the appearance of the snails in certain areas results from the action of predators hunting by sight. The predators include small mammals, such as the shrew, which may select by tone as most of the mammalian species are color blind. Birds, especially the song thrush, are the chief enemies of the land snail, as they can distinguish color, and they smash shells on convenient stones, 'thrush anvils', to obtain the soft body. By studying the shell remains it can be decided whether there is any color selection by the thrush, or whether it eats snails at random. Much study in different areas has shown that the birds capture snails selectively, destroying an unduly large proportion of those whose color and patterns least match their habitat. Here again natural selection is in action, although in certain areas selection for color pattern appears absent due to the wide range of specimens found.

70

Thrushes are the main enemies of the land snail.

A selection of shell patterns of *Cepaea nemoralis*

COURSE OF EVOLUTION

The long period from the origin of the Earth, about five billion years ago, to the beginning of the Cambrian period, 520 million years ago, is often referred to as the barren past. However, the pre-Cambrian period definitely supported certain unspecialized principal forms of life, as the Earth probably became habitable some three billion years ago. Very few fossils can be found from the deeper pre-Cambrian rocks, although the Cambrian rocks reveal a wide and diversified account of animal and plant life.

Several plausible explanations are available, however, to account for this lack of fossil evidence. Many of the fossil forms may have been destroyed by erosion or by metamorphosis during the passage of time, or they may be as yet undiscovered. However, an important point is that the early forms were probably soft bodied, and therefore incapable of preservation as fossils.

The earliest fossils known to us are still extraordinarily complex in comparison with the lowly organisms which must have existed at the dawn of life. Pre-Cambrian fossils tend to cause

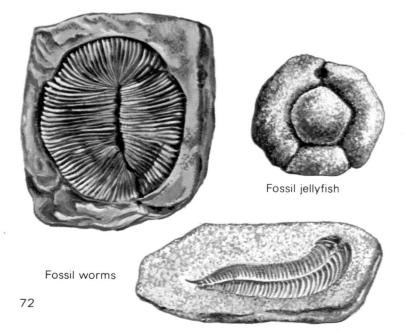

Fossil jellyfish

Fossil worms

A middle-Cambrian scene

most confusion to the paleontologists regarding their age and their actual form. A medusoid impression that probably represents a jellyfish has been discovered in pre-Cambrian rocks of the Grand Canyon, some algae or sponges have been identified from Scotland's pre-Cambrian rocks, and important evidence of worms, coelenterates and echinoderms has come from Australia.

One often hears of the large gap between the pre-Cambrian period and the lower-Cambrian period when apparently there was a sudden appearance of a wide range of well-fossilized plant and animal life. It must be remembered, however, that the Cambrian period lasted some some 90 million years, so there was time enough for apparent 'sudden' forms to occur. There are abundant records of marine invertebrates and algae, but trilobites appear to have been the dominant group. In fact, there may have existed, even at this time, representatives from all the phyla of plants and animals.

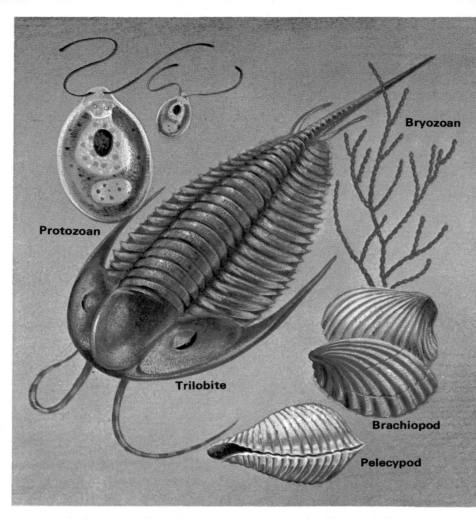

A diverse number of invertebrate forms evolved in the shallow seas of the Paleozoic era.

Marine invertebrates

One must remember that the earth was far different at the beginning of the Paleozoic era than it is today. An enormous land mass appears to have stretched then from the Galapagos Islands, west of South America, to western Australia. The climate was equable, but the land was barren except for patches of green algae encrusting the rocks. In the seas, life had not evolved beyond invertebrate fauna. However, even among these animals there was tremendous radiation into a diverse number of forms.

Present in the shallow seas were sponges, jellyfish, gastropods, brachiopods, worms, arthropods and echinoderms. The sponges (phylum Porifera) are the simplest of all the multicellular animals. The Cambrian sponges were fixed in position and varied widely in size, shape and color. Until 100 years ago their actual place in the classification table was undecided. They were finally shown to be animals, probably arising from protozoans. The body of a sponge is more or less a sieve made up of animal cells. The body wall is perforated throughout by pores and canals, and water flows into the central cavity and out through the top by one or more openings. Many sponges are supported by silica spicules, either separate or joined to form a skeletal framework. Often these have been fossilized, producing beautiful patterns.

Above these fixed dwellers on the sea floor floated the jellyfish polyps. Their soft bodies were rarely preserved, but even so they are known from early Cambrian times. Sea anemones, corals, sea fans, and sea pens are all grouped with the jellyfish in the phylum Coelenterata. All are aquatic, either solitary or colonial, and have a sac-like body cavity with a ring of tentacles surrounding the single opening. They feed on a wide range of life, some capturing and eating only microscopic organisms but other more formidable ones devouring mollusks, crustaceans, and fish which are first paralyzed by means of stinging cells.

The most characteristic of all forms of life in the Cambrian seas were the brachiopods, the bottom-living marine 'shellfish' and the ruling trilobites. Today, there are only about 200 species of brachiopods alive in the seas, but in the lower Paleozoic they were very abundant and universal. The brachiopod shell consists of two valves, bilaterally symmetrical, surrounding and protecting the soft parts of the body. Instead of the two valves matching each other, they differ in size, unlike a true bivalve shellfish. The animal is usually attached to the ocean floor by a long, fleshy stalk or pedicel, which usually perforates the larger valve. *Lingula,* a small tongue-shaped species, is an example of a simple brachiopod present in Cambrian seas that has remained unchanged to this day.

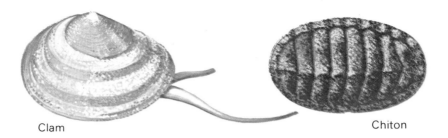

Clam

Chiton

Mollusks

Slow-moving snails found in the rock pools of Cambrian times were the oldest representatives of one of the most varied and successful groups in the world today. The phylum Mollusca includes animals such as slugs, snails, oysters, clams, chitons, cuttlefish and octopuses. There is a great variety in size, form and shape, but all share a fundamentally similar body plan. They have a soft unsegmented body, an anterior head and a large visceral mass supported on a fleshy muscular foot. Surrounding most of the body is a thin fleshy layer, the mantle, which secretes a shell in most members of the phylum. Snails have a single coiled shell, while in the clams and oysters the shell consists of two valves. The chiton has eight protective movable plates, while in cuttlefish and squid the shell is completely internal.

Although the early forms were entirely marine, as conditions changed the gastropods slowly evolved into a most widely adapted and diverse group. Today the 20,000 modern species inhabit a very wide range of environments: marine, freshwater, terrestrial, arctic, tropic, desert and from ocean depths to over 18,000 feet above sea level.

Members of the phylum Mollusca (*top* and *bottom*) and the evolution of various invertebrates (*opposite*)

Snail

Octopus

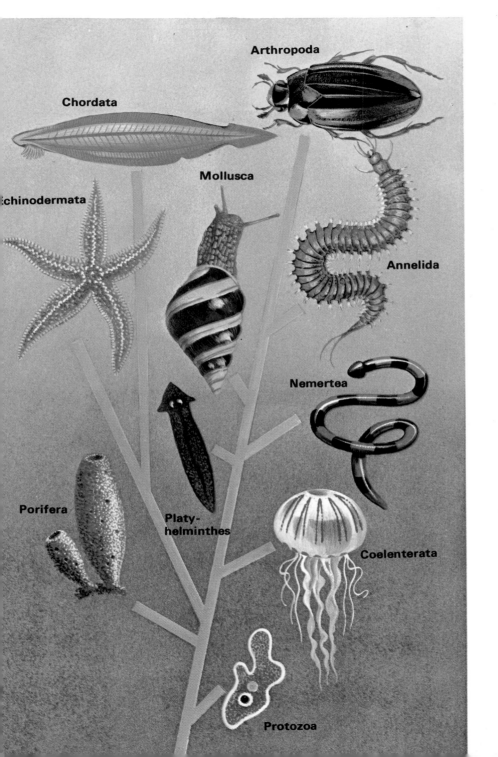

Arthropoda

Chordata

Mollusca

Echinodermata

Annelida

Nemertea

Porifera

Platy-
helminthes

Coelenterata

Protozoa

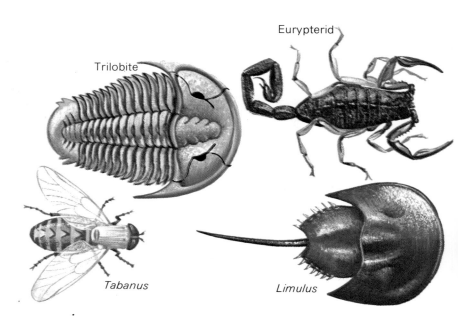

Trilobite

Eurypterid

Tabanus

Limulus

Primitive (*top*) and more modern representatives (*bottom*) of the phylum Arthropoda

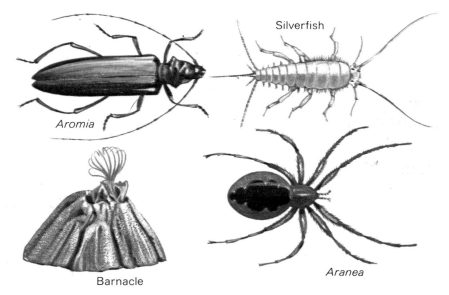

Aromia

Silverfish

Barnacle

Aranea

Arthropods

The Arthropoda contains about three-quarters of all known animal species including shrimps, crabs, lobsters, barnacles, spiders, scorpians, ticks and insects as well as the extinct trilobites and eurypterids.

Dominating the Cambrian seas were the trilobites, and although some 1,000 genera were alive in that period, today they are completely extinct. The woodlice-like appearance is perhaps one of the most familiar forms of all fossils. Trilobites underwent adaptive radiation and evolved many forms suited to particular environments. They started a steady decline in the Ordovician until they had dwindled to extinction in the late Permian times, some 225 million years ago.

The extinct eurypterids are the ancient aquatic relatives of the scorpions. Their usual length was about one foot, although some reached nine feet in length. Scorpion-shaped, the oldest eurypterids originated in the Ordovician, reached their peak in the Silurian and the Devonian and became extinct in the Permian.

Also related to the eurypterids are the horseshoe crabs, still living today, whose earliest ancestors appeared in the Cambrian. *Limulus,* the king crab, is often spoken of as a living fossil, and although only marine forms occur today, they were the first animals to colonize fresh water.

The complete animal colonization of fresh water and land could take place only after colonization by plants. It is therefore no accident that terrestrial animals and plants appear at roughly the same period of geological time. Plants provided the food for the first terrestrial animals, which were arthropods. The interdependence of plants and animals is of great evolutionary importance in the appearance of certain groups of animals and, as the flora of the various periods changed, in the decline of others.

It is amazing how successful the orders of the arthropods have been in the conquest of the land, and certain characteristics are responsible for this. Extremely important is the chitinous shell which protects them from the hazards of desiccation. Furthermore, arthropods were already equipped with appendages which could support their bodies when out of the water, as well as provide an efficient means of locomotion.

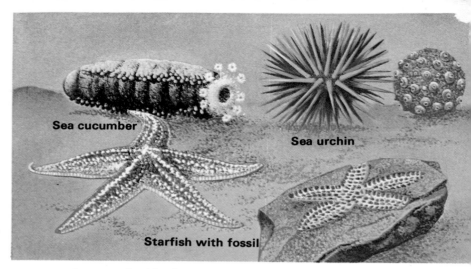

A group of echinoderms

Echinoderms, annelids and bryozoans

The phylum Echinodermata includes the living starfish, sea urchins, brittle stars, sea lilies and sea cucumbers, as well as the extinct blastoids, cystoids and edrioasteroids. The earliest forms date back to the Cambrian period, and the phylum is entirely marine although some are free-living while others are sedentary. They all tend to be spiny, and the body is encased either by a calcareous roundish test as in the sea urchins, or by a leathery skin having a star shape and studded with small calcareous plates as in the starfish. Usually a five-fold symmetry is apparent. The larval forms show a striking resemblance to the larvae of hemichordates, and this has led to the linking of the two groups by some zoologists. The echinoderms may be close to the forms from which the chordates evolved.

The Cambrian echinoderms are all representatives of archaic and now extinct groups. *Edrioaster,* an extinct edrioasteroid, was globular, with five sinuous arms radiating from the mouth. Eocrinoids living attached to the sea floor in the Cambrian were probably ancestors to the later crinoids and cystoids. By the middle Paleozoic, the familiar starfish, sea urchins, sea cucumbers and crinoids, as well as a number of now extinct forms had evolved. The most important of the Paleozoic echinoderms were the crinoids and the extinct blastoids, both of which first appeared in the Ordovician. Crinoids have been common in limestone deposits from the Silurian onward.

Some were free-swimming, others were fixed by a stem.

Included in the phylum Annelida are the marine bristle worms, the Polychaetes, the most generalized class of annelids. The earthworms and their relatives are grouped in the Oligochaeta (few bristles). The leeches, class Hirudinea, are also annelids. In Cambrian times only marine worms had evolved, but they seem to have been as widespread as the marine worms of today. The soft bodies were not suitable for fossil preservation, but nevertheless evidence of their form and habit is often found. The worms are usually represented by fossil tracks, tubes, burrows and parts of their chitinous jaws (scolenodonts).

Bryozoans were not common until the Ordovician period, and geologists seem to have neglected this group. Many bryozoans resemble corals in the external form as the individual animal secretes a calcareous mat-like or frond-like skeleton. Bryozoa is a name meaning 'moss animals' and refers to the plant-like appearance of many bryozoans.

Representatives of the Annelida and Bryozoa

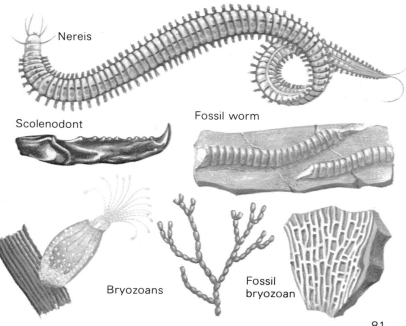

Nereis

Scolenodont

Fossil worm

Bryozoans

Fossil bryozoan

Origin of the vertebrates

During the Silurian there were equable climates and a number of large land masses surrounded by warmish seas, where limestone, sand and mud could slowly accumulate. The invertebrates were still masters of the sea, but it was during these times that the vertebrates and land plants emerged.

A vertebrate has a hard, narrow supporting structure extending along the dorsal surface of its body called a backbone. In most characteristic groups this is made of true bone and segmented to give rigidity and flexibility. We classify animals with this backbone into the phylum Chordata.

The evidence for the evolution of the vertebrates comes from the study of embryos and by comparative anatomy. Clues are found by investigating a few peculiar vertebrates which at first sight bear little resemblance to the higher backboned species. They are known as the Protochordata, which includes the sea squirts or ascidians, amphioxus and acorn worms. They are classed with the vertebrates because they possess, at some time in their life history, a stiffening rod (the notochord), pharyngeal gill pouches or slits, and a dorsal tubular nerve cord—all vertebrate characteristics.

The tunicates or sea squirts in their adult form seem to be unrelated to the vertebrates. They are sedentary, sac-shaped and almost sponge-like, with a stiff outer body, often colonial, clinging to rocks and seaweeds in shallow seas. They do not have a nerve cord, vertebrae or a notochord, although they have gill slits. However, the sea squirt is believed to be a vertebrate, and the evidence comes from the larval form. The larva is free-swimming and tadpole shaped, with a notochord and a nerve cord.

The sea squirt larva closely resembles the adult amphioxus, the lancelet. Amphioxus is another early chordate, about two inches in length, translucent and streamlined in shape. It, too, has a notochord, a dorsal nerve cord and gills with a simple ventral digestive tract.

Although these modern chordates have undoubtedly changed from the original unspecialized forms, they are not very different from the early chordates, which provide a link between the invertebrates and the vertebrates.

Protochordates provide clues to the origin of the vertebrate form.

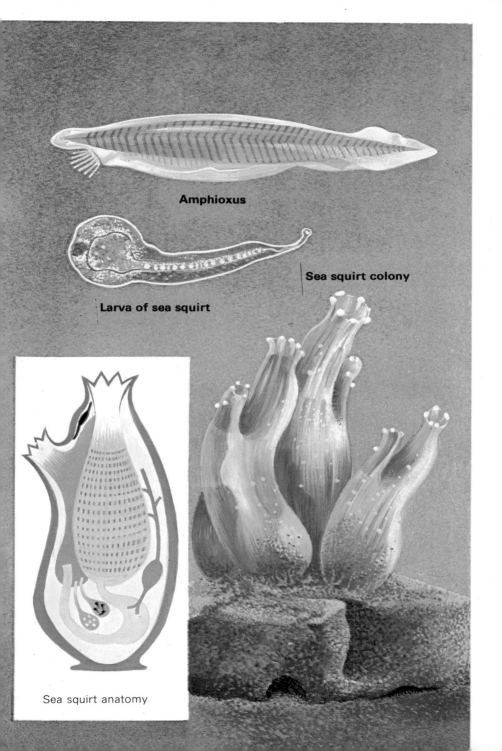

Amphioxus

Larva of sea squirt

Sea squirt colony

Sea squirt anatomy

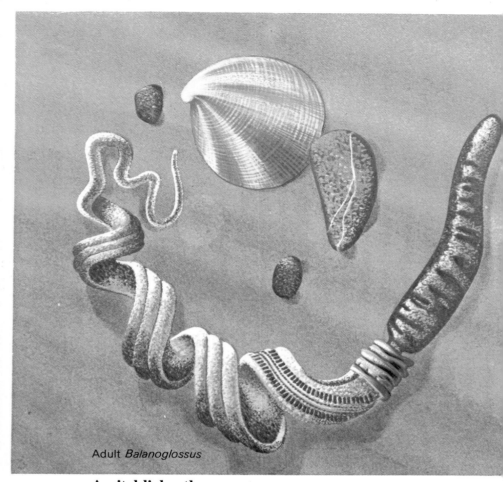

Adult *Balanoglossus*

A vital link—the acorn worms

The acorn worms are worm-like, simply organized burrowing marine worms that include *Balanoglossus, Glossobalanus, Ptychodera* and *Saccoglossus*. They are the most primitive of vertebrates and in external appearance resemble invertebrates rather than vertebrates. Found in burrows in the sands and muds of shallow marine waters, they feed by extending the mouth from their burrow, and drawing particles of sand and mud down the simple alimentary canal which then extracts the organic matter. The mouth is not to be mistaken for the proboscis, a muscular, cylindrical burrowing structure which retracts into a collar surrounding the mouth.

The identifying chordate structures are a dorsal nerve cord

and a shorter ventral one, gill slits and a notochord. Again it is the larvae that are of extreme importance. They are small, bell-shaped organisms which drift with the currents of the sea. Around the mouth is a ring of cilia which changes shape as the larva develops. The larval form is termed a *tornaria* larva. This primitive chordate larva when compared with the larvae of worms and mollusks shows similarity, but only the mollusks have a single ciliated ring encircling the body in front of the mouth. It is feasible that larvae such as these may have given rise to the later vertebrates. However, by which route and from which invertebrate group the vertebrates evolved needs further investigation. Certain authorities state that arthropods and annelids provide the link because these animals have a gut and nerve cord. However, the gut in the arthropods is dorsal and the nerve cord is ventral, so to produce a characteristic vertebrate organization, they must be reversed and a complete inversion of two complex parts seems unlikely. The current view is that the most likely ancestors of chordates are the echinoderms. Here again it is the larval echinoderm which links the invertebrates quite definitely with the lower chordates. The larvae of echinoderms closely resemble the larvae of the acorn worms.

Research has also shown physiological and anatomical similarities. The evidence points to a common ancestor for the echinoderms and chordates which probably became extinct and for which no fossil evidence has yet been found.

Larvae of an echinoderm (*left*) and *Balanoglossus* (*right*)

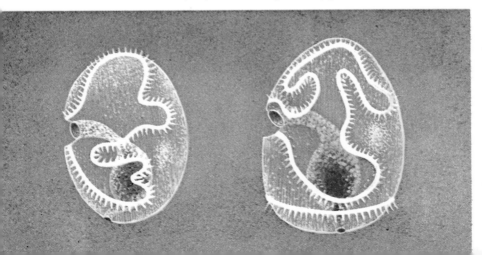

Coccosteus

Pterichthys

Cephalaspis

The earliest vertebrates

The evolution of some vertebrate larval forms to primitive fish-like forms can hardly be expected to have left traces in rock, as they both were most probably soft-bodied. The earliest traces are from the middle Ordovician period, and some body fragments have been discovered in the Colorado and Wyoming area. There are four classes of fish identified today, all having gills, skin with scales and fins: the Agnatha, the Placoderms, the Chondrichthyes and the Osteichthyes. All had evolved by Devonian times. The most primitive are the Agnatha, which are the only craniates lacking true jaws and paired fins. The Devonian period saw a great expansion of the fish, and much fossil evidence has been discovered so that good coverage of the groups that had evolved is available.

A fish typical of the Agnatha group and found in the Devonian is *Cephalaspis*. As with most cephalaspids it had a flattened, bony head shield and a scale-covered body. Cephalaspids appear to have marine and freshwater members,

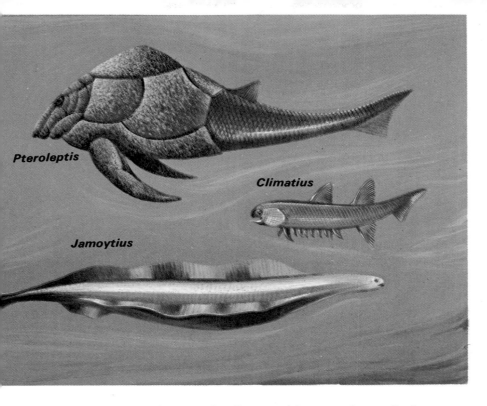

Pteroleptis

Climatius

Jamoytius

probably all were bottom dwellers, grubbing in the mud of streams and ocean bottoms. *Jamoytius* is torpedo-shaped and bears median lateral fin-folds, with large eyes. Superficially it is not unlike amphioxus.

The Placoderms appeared in the upper Silurian. They had developed primitive jaws and paired fins and radiated into a variety of forms. The development of jaws and fins are of extreme importance and represent a landmark in the evolution of all the vertebrates, not only among the higher fish. Until the appearance of jaws, animals were confined to the mud-grubbing life, but with jaws new food sources became available together with new aquatic environments. Without bony fins, animals could never have left the water for a terrestrial life.

The Placoderms probably originated from primitive Agnatha and although the whole group marked a major step in the evolution of the vertebrates, they were reduced by the end of the Devonian and extinct by the close of the Paleozoic.

87

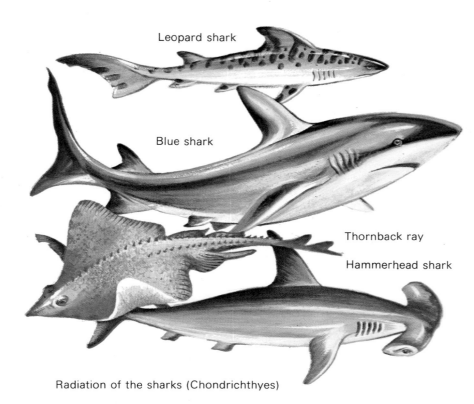

Leopard shark

Blue shark

Thornback ray

Hammerhead shark

Radiation of the sharks (Chondrichthyes)

Cartilaginous fish—Chondrichthyes

Sharks, rays and chimeras are classed in the Chondrichthyes. They are cartilaginous, predatory fish and usually powerful swimmers. They depend on their large nostrils for tracking down their prey, rather than on their eyes which are quite small. The sharks appeared in the Devonian and underwent considerable radiation, becoming adapted to all types of habit by the upper Paleozoic. Many became extinct by the end of the Permian but even so, today, there are about 600 successful species distributed widely throughout the oceans of the world. Fossil evidence shows that the early representatives were remarkably like the modern living species. *Cladoselache,* a well-known fossil from the upper Devonian of Ohio, is some three feet in length with well-developed, broad fins and a streamlined naked body.

Bony fish—Osteichthyes

Bony fish are the most abundant, diverse and complex group of fish. They appeared in mid-Devonian times and flourished greatly throughout in the rest of the Paleozoic. By this time they were almost in sole possession of freshwater lakes and streams. They have a bony skeleton and scale-covered bodies. Some fossil fish and a few living fish have lungs; the rest have a swim-bladder which controls buoyancy.

From the outset there appears to have been two distinct types of bony fish. The first of these are the Actinopterygii, or ray-finned fish, the four orders of which are the palaeoniscids, the chondrosteans, the holosteans and the teleosts. Most have died out, but the teleosts are dominant today.

The second division of the bony fish is Sarcopterygii, a group containing air-breathing fish. This group contains lobe-fins (including coelacanths) and the dipnoans, or lungfish. The importance of this group lies in their close relationship with the ancestors of the higher vertebrates (see page 98).

Examples of the diversity found among bony fish today.

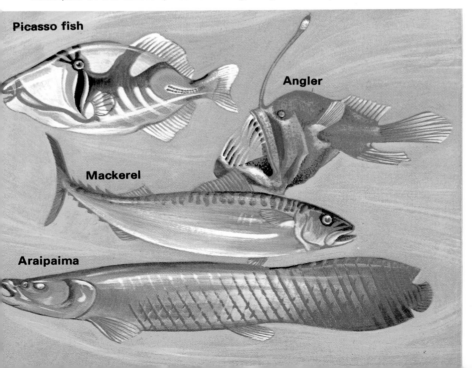

Picasso fish

Angler

Mackerel

Araipaima

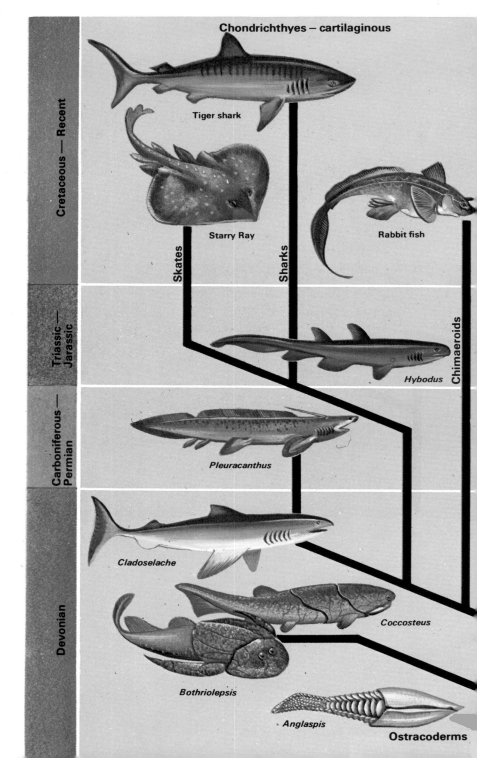

Chondrichthyes – cartilaginous

Cretaceous — Recent

Tiger shark

Starry Ray

Rabbit fish

Skates

Sharks

Chimaeroids

Triassic — Jurassic

Hybodus

Carboniferous — Permian

Pleuracanthus

Devonian

Cladoselache

Coccosteus

Bothriolepsis

Anglaspis

Ostracoderms

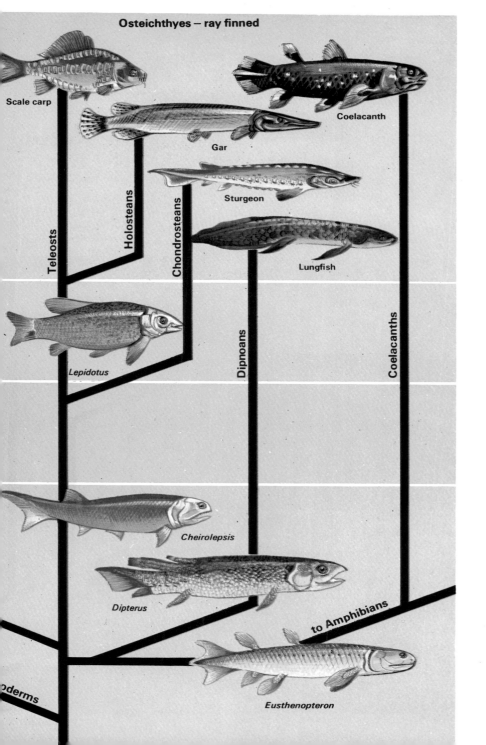

Osteichthyes – ray finned

Scale carp

Coelacanth

Gar

Teleosts

Holosteans

Chondrosteans

Sturgeon

Lungfish

Dipnoans

Coelacanths

Lepidotus

Cheirolepsis

to Amphibians

Dipterus

oderms

Eusthenopteron

Polypterus

Actinopterygii — ray-finned fishes

The Actinopterygii fins are composed of a web of skin supported only by horny rays. Flesh and bone is confined to the base of the fin. Many marked differences distinguish this group from the Sarcopterygii, including scale structure, different patterns of the bones of the head and lack of internal nostrils. The eyes tend to be large and seem to be the dominant sense organ; the sense of smell is of relative unimportance. The lung is transformed into a hydrostatic organ, the swim-bladder. The number of eggs laid is far higher than any other fish group, and although small in size they number up to thousands and even millions from one female alone. Quantity and not quality is the principle and although many may die in early life, the numbers are quickly renewed, probably an important factor in the success of the group.

Chondrosteans

The chondrosteans, or palaeoniscids, are the oldest ray-finned fish. They first appeared during the mid-Devonian. At that time, they were greatly out-numbered by the lung-fish and the lobe-finned fish, both members of the Sarcopterygii.

Bowfin

Sturgeon

Polypterus, a present-day chondrostean found in central Africa, is a modified form of its early Paleozoic ancestor *Palaeoniscus.*

Two other members of the chondrostean group are sturgeons and paddlefish. Both are found in North America. These fish have a number of rather degenerate features. They have very few scales and their once-bony skeleton is now mostly cartilage. They also have relatively weak jaws and scavenge for food at lake and stream bottoms.

Holosteans

The holosteans emerged in Permian times from the palaeoniscid stock, which they largely replaced. They had deep bodies and a more advanced symmetrical homocercal tail. The only survivors today are found in the lakes and streams of North America. They are completely successful in spite of the competition from higher bony fish. The bowfin *(Amia),* or freshwater dogfish, and gar-pike *(Lepisosteus)* differ quite markedly in external features.

Garpike

Dragon fish

Sea Horse

The modern dominant teleosts

Arising during the late Triassic from holostean stock were the teleosts. They were comparatively rare until the Cretaceous, by which time several lines of evolution had already begun. Today, some 20,000 bony fish species abound in the seas and waters illustrating the success of the group.

The teleost plan of evolution has allowed for the development of a great range of specializations, fitting the fish to all sorts of situations in sea and fresh water. Since sea and fresh water have been in existence relatively unchanged throughout the course of the evolution of fish, one might ask why there was a need for the fish to evolve new forms to suit apparently 'new' habitats. In a sense, the fish did not evolve to fit a new environment, but rather they found endless new ways of

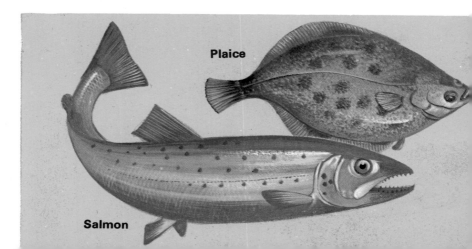

Plaice

Salmon

Goldfish

Archer fish

living in the same aquatic environment.

The teleosts vary greatly in every respect. The more primitive body shape and general organization is shown by the herring and sardine, with slightly more advanced forms shown by the trout and salmon. Fin rays are more numerous and flexible in the primitive types, whereas later the fins are supported by only a few stout, movable spines. The body form takes on such tremendous variation that it seems in some comparisons that the fish concerned could not possibly be related. The flatfish, such as the plaice, sole and flounder, have evolved a form perfectly adapted for life on the sea floor. They all have very flattened bodies, and during development the young fish turns itself through 90° and the lower eye migrates to the new upper surface.

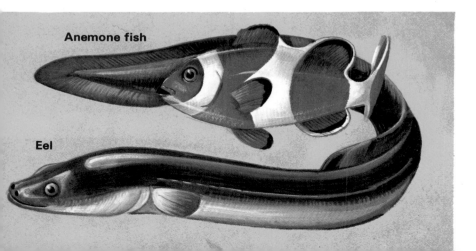

Anemone fish

Eel

The coelacanth

The Sarcopterygii are a group which became rare in the Carboniferous and apparently disappeared after the Permian. Knowledge of these lobe-finned fish was quite limited as no members had apparently survived. From fossils it was evident that they had lungs and internal nostrils and in many respects resembled the primitive amphibian skeleton. The fish lacked legs, of course, but their fins contained a fleshy lobe, within which were bony skeletal supports. The basic pattern of these fish is therefore comparable to that of a land vertebrate. The early typical lobe-fins lived in the freshwater areas and become extinct quite early. No later fossils were known, so it was assumed that the sarcopterygians were entirely extinct.

However, the amazing discovery in 1939 of a coelacanth fish off East London, South Africa, gave evidence that for more than 70 million years coelacanths had survived comparatively unchanged. Several specimens have now been caught, so more extensive knowledge of the fish is available.

The 'living fossil' coelacanth

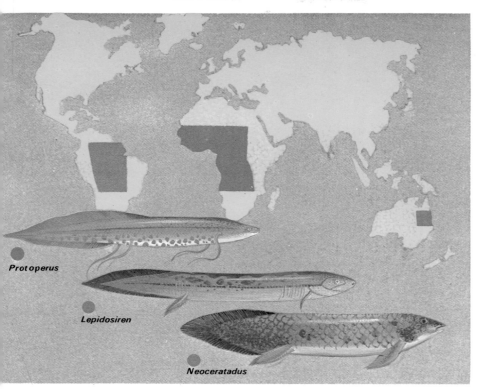

Distribution of the lungfish

The lungfish

Only three genera of lungfish, also called dipnoans and sarcopterygians, survive in the tropics: *Neoceratodus* in the Burnet rivers of Queensland, Australia; *Protopterus* in the White Nile, some of the great lakes and the Congo region of Africa; and *Lepidosiren* in the Amazon and Parana rivers of South America.

The lungfish possess some of the features that enabled some of their evolving relatives to emerge from the water and become terrestrial vertebrates. The critical evolutionary period took place in stagnant swamps and drying pools, where the waters were low in oxygen and subject to periodic evaporation. The lungfish, because they possessed lung structures, became proficient in gulping down oxygen directly from the air at the surface. The early ancestors of lungfish may well have branched to give rise to amphibians, and indeed the larvae of *Lepidosiren* and *Protopterus* shows similarity to those of amphibians by having suckers and external gills.

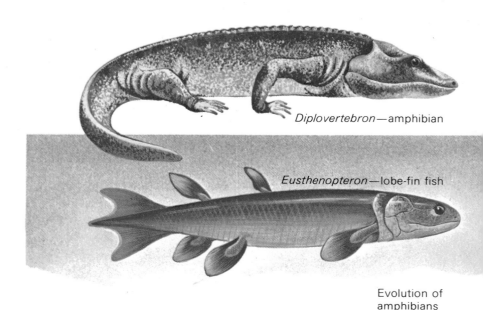

Diplovertebron—amphibian

Eusthenopteron—lobe-fin fish

Evolution of
amphibians
(*opposite*)

The conquest of land

The change from aquatic to terrestrial life by some of the
vertebrates was first initiated in the Devonian by the early
amphibians and completed by the reptiles later in the Paleozoic. Once more it was not a rapid process, but a slow progres-
sion through various evolutionary stages. Certain bony fish had
already developed essentials for land life, for example, some
had developed primitive lungs. Furthermore, an active
life on land for any vertebrate demands that there is good
support for the body. In the lobe-finned fish (e.g., *Eusthenopteron*) the stout and fleshy lobe in the paired fins gave
possibilities for development into land limbs. Many of the
requirements necessary for terrestrial life were already initiated
in the bony fish.

Comparing the skulls of the earliest amphibians and those of
the Devonian lobe-fins Crossopterygii fish, there can be little
doubt of the relationship between the amphibians and some of
these fish. However, intermediate fossils have still not been
discovered. Something of the transition is revealed by the
oldest fossil amphibia, the ichthyostegids from the upper
Devonian of East Greenland, which show features intermediate between lobe-fins and the typical early Amphibia.

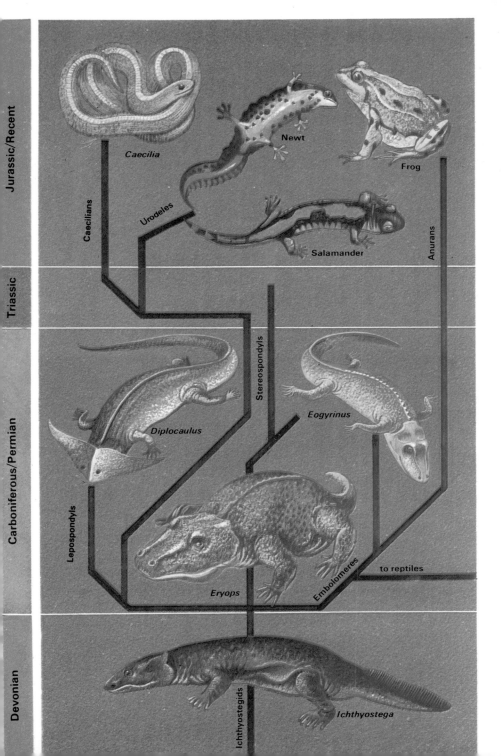

Jurassic/Recent

Triassic

Carboniferous/Permian

Devonian

Caecilia

Newt

Frog

Caecilians

Urodeles

Salamander

Anurans

Stereospondyls

Eogyrinus

Diplocaulus

Lepospondyls

Eryops

Embolomeres

to reptiles

Ichthyostegids

Ichthyostega

Eryops, about 5 feet long

Amphibians past and present

Early amphibian fossils are rare, but excellent skulls and other skeletal parts from the late Devonian have been found in Greenland. These amphibians of the Devonian and Mississippian periods, with their large heads, solidly roofed with thick bones, are stegocephalians, and show little resemblance to frogs and salamanders. The most conspicuous animals of the Pennsylvanian were the amphibians. They had strong legs and were able to move on land as well as in water. The giants in the water were the embolomeres; their limbs were reduced, but they had powerful tails. *Eryops,* a large labyrin-

Salamander-like fossil larvae branchiosaurs

thodont was found by lakes and streams during the late Pennsylvanian and Permian. Although reptiles were becoming increasingly dominant these formidable amphibians were quite able to compete with them. Some developed armor which allowed them to leave the water without dehydrating.

The amphibians we know today—frogs, newts and salamanders—did not evolve until the Jurassic. Salamanders are similar in general appearance to the ancient types, but in the skull and skeleton there are many degenerate features. Their scaleless, moist skin acts as an accessory breathing organ and to avoid desiccation they must remain in moist areas.

The tailless frogs are a flourishing group. Although they are often taken as examples of primitive amphibians, they are among the most specialized of vertebrates and this is attributable mainly to their hopping method of locomotion. They are

Salamander   Frog

well adapted for movement both on land and in water and can breathe in both environments providing they keep to moist areas when on land. They are also dependent on water for breeding and spend the cold winter months in hibernation. The frogs, therefore, well illustrate both the advances of the amphibians and the limitations of the amphibian way of life.

The Apoda, or limbless amphibians, are inconspicuous burrowing animals, much like worms, found in the tropics. Their fossil history is unknown, but they are a degenerate group, not closely related to the other orders.

Seymouria, about 2 feet long

Reptiles

The amphibians, although the first to conquer the land, are a defeated group. Abundant at first, they are now an insignificant group among the tetrapod vertebrates. The reptiles have truly won the land, and the reason is quite straightforward when the development of the two groups is studied. Amphibians throughout their life are dependent on water; their eggs are either laid in water or in a very moist environment, to which the adult must periodically return for reproductive purposes.

With reptiles, no aquatic stage is necessary as they have evolved a type of egg which can be laid on land. The amphibian embryo, developing in water, derives its oxygen, food and protection from its watery surroundings. The evolution of the reptile's egg (called amniotic egg) provided the same benefits on land or within the mother's body, and the development of copulatory organs allowed internal fertilization.

From the early evolved reptiles, there branched numerous kinds, two paths evolving to give rise eventually to the warm-blooded birds and mammals. The early reptiles are known as cotylosaurs and are referred to as the stem reptiles. Although the distinctive characteristic is the amniotic egg, there are many other ways in which the reptiles differ from the amphibia. Many parts of the skeleton show differing structures and also circulatory and respiratory systems. The cotylosaurs probably developed from the seymouriamorphs. *Seymouria,* a squat creature about 2 feet in length, from the lower Permian of Texas, has given its name to the group. However, it is rather a misleading creature in its characteristics, some of which are amphibian, others reptilian. The structure of its skull, teeth and some features of its vertebrae are typically amphibian. Other characteristics of the vertebrae are reptilian, as is the jaw suspension, the structure of the shoulder girdle and the limbs. *Seymouria* itself is too recent to be the direct ancestor of the reptiles, but it is probably not greatly different from its ancestors. The question of which group it belonged to would soon be solved if it was known whether it laid eggs on land or in the water. This does emphasize that it is the change in development which is the important feature in the history of reptiles.

Dimetrodon,
about 11 feet
long

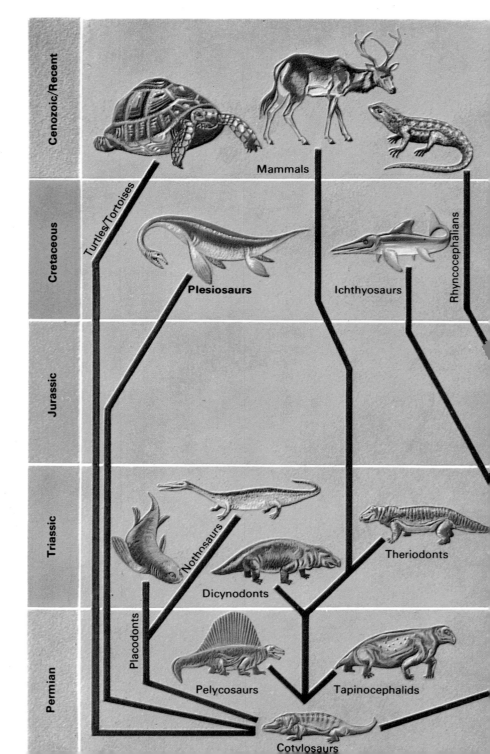

Cenozoic/Recent

Cretaceous

Jurassic

Triassic

Permian

Turtles/Tortoises

Mammals

Rhyncocephalians

Plesiosaurs

Ichthyosaurs

Nothosaurs

Theriodonts

Placodonts

Dicynodonts

Pelycosaurs

Tapinocephalids

Cotylosaurs

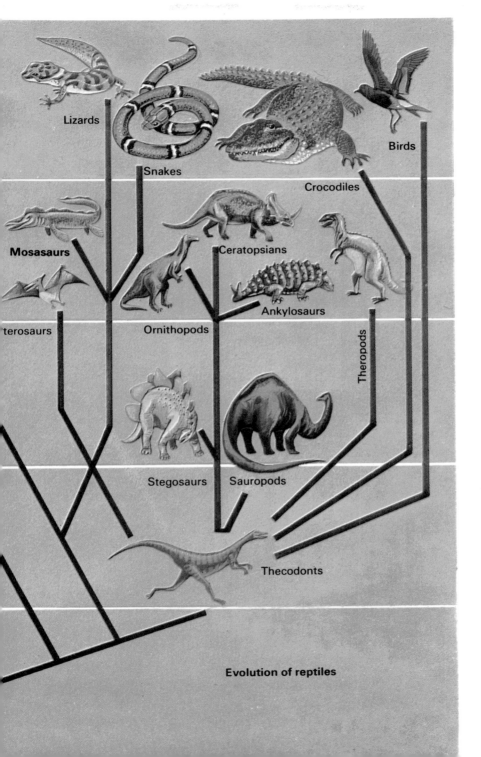

Lizards

Snakes

Crocodiles

Birds

Mosasaurs

Ceratopsians

terosaurs

Ankylosaurs

Ornithopods

Theropods

Stegosaurs

Sauropods

Thecodonts

Evolution of reptiles

A reptilian Jurassic scene

The age of the reptiles

The age of the reptiles refers to the Mesozoic era, which was made up of the Triassic, Jurassic and Cretaceous periods, lasting some 165 million years. Birds, mammals and modern insects appeared for the first time during this period, but the reptiles were the dominant group.

Early Mesozoic reptiles included a great variety of forms, some giving rise to the early ancestors of the ichthyosaurs and plesiosaurs which moved back to the sea. Among the terrestrial reptiles, certain ones became bipedal, light-boned and fast-running saurians and gave rise to the dinosaurs. The cotylosaurs were the stem reptiles from which these later advanced groups radiated. The pelycosaurs evolved in the late Carboniferous and were destined to play a most important part in the evolution of higher vertebrates. The pelycosaurs were a varied group, but most were sprawling creatures with long tails. Some exceeded 10 feet in length, and the most striking parts of these reptiles were their backs, which had greatly elongated neural spines, which appear to have been covered by skin thereby forming a sail-like structure. *Dimetrodon* was around 11 feet long, a quite active predator with a large, deep skull and differentiated teeth. Its sail fin was about 3 feet high, but the function of this curious structure still remains a mystery. It may have been concerned with temperature control or may even have been a form of protection.

It was from the pelycosaurs that the other great synapsid group, the therapsids, developed. This group became widespread in the Permian and Triassic. Some members are strikingly similar to the mammals, which descended from them.

The dicynodonts (dog-toothed) were a common and widespread Permo-Triassic group of large vegetarians with reduced teeth. In the theriodonts (beast-toothed), the mammalian resemblances are most striking. *Cynognathus,* an active carnivore, is a typical member of this group.

Tiny and even more mammalian in character were the ictidosaurs, or weasel lizards, of the upper Triassic. The jaw articulation in some is of the mammalian type.

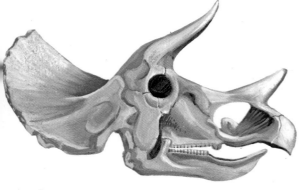

Fossil horned dinosaur skull, tracks and eggs

The dinosaurs

Although the Cretaceous saw the first appearance of mammals, it was the time of the mighty dinosaurs. The popular belief about dinosaurs is that they comprised a single group of gigantic reptiles. However, although many reached 50 tons, some were the size of hens. Also, they were not a single group, but two distinct orders which had divided from the primitive stock. Early members of both orders were bipedal but several forms later became tetrapods.

Members of the order Saurischia are identified by the structure of the pelvic girdle. The early members, such as *Compsognathus* and *Ornitholestes,* were small and active bipedal runners. The positioning of the toes was similar to those of perching birds, three forward and one backward. This led to a false identification of the dinosaur footprints discovered in Triassic rocks, as they were first thought to be prints of primitive giant birds. By the end of the Cretaceous the Saurischia had produced the largest carnivores that have ever lived, such

Cretaceous dinosaurs

as *Tyrannosaurus,* measuring nearly 50 feet long and 20 feet high. They probably preyed upon the herbivorous dinosaurs and both became extinct, either because of climatic changes or competition with the mammals and birds.

An offshoot from this line gave some ostrich-like forms, *Struthiomimus* and *Ornithomimus,* which walked on three toes with a grasping three-fingered hand. They were toothless, but developed a horny beak and probably ate eggs. Another line gave rise to the large four-legged herbivores: *Brontosaurus, Diplodocus* and *Brachiosaurus.* These were the largest of all terrestrial animals. *Brachiosaurus,* found in North American and East African deposits, was the real giant—some 80 feet long and weighing close to 50 tons—but with quite a small tail.

The second order, Ornithischia, included the duck-billed dinosaurs, which often had 2,000 teeth and a beak. Most were heavily armed—*Stegosaurus* had spines on the back and *Triceratops* had huge horns and a bony neck shield.

Marine reptiles

The reptiles were the first group to firmly establish themselves on land, but they evolved some members which returned to the sea. Often when a particular group is well established in one habitat, it will also spread into other environments.

The reptiles that returned to the seas re-adapted to aquatic conditions and became as diverse as the shore reptiles. Six groups have been identified.

The turtles appeared in the Triassic and are little changed. They seem to have evolved from a reptile similar to *Eunotosaurus*. Land tortoises did not evolve until the Tertiary with the freshwater terrapins and marine turtles radiating from them. In the Cretaceous some marine turtles grew to a length of 12 feet.

Highly adapted to an aquatic life were the ichthyosaurs or fish-reptiles, so called because of their very fish-like appearance. Their limbs were highly modified and they could there-

Marine fossil reptile fauna

fore not come onto land to breed. It is suggested that the ichthyosaurs retained their eggs inside the body until they hatched,
as do some snakes and lizards. Young ichthyosaurs have been
found within the body of an adult fossil specimen. They must
have swam in a fish-like manner, using the tail as the main
propulsive force. The oldest fossils are from the mid-Triassic
and they continue until the late Cretaceous.

The most savage reptiles were the mosasaurs and geosaurs.
The geosaurs were marine crocodiles of the Jurassic and lower
Cretaceous. The mosasaurs seem to be confined to the Cretaceous. Some grew to be 30 feet long. Many of these aquatic
reptiles had long snouts and sharp teeth, which were used for
eating fish. *Placodus*, whose skull is shown on the next page,
had large flat teeth for crushing shellfish. Plesiosaurs, another
group of aquatic reptiles found throughout the Jurassic and
Cretaceous, grew up to 50 feet long.

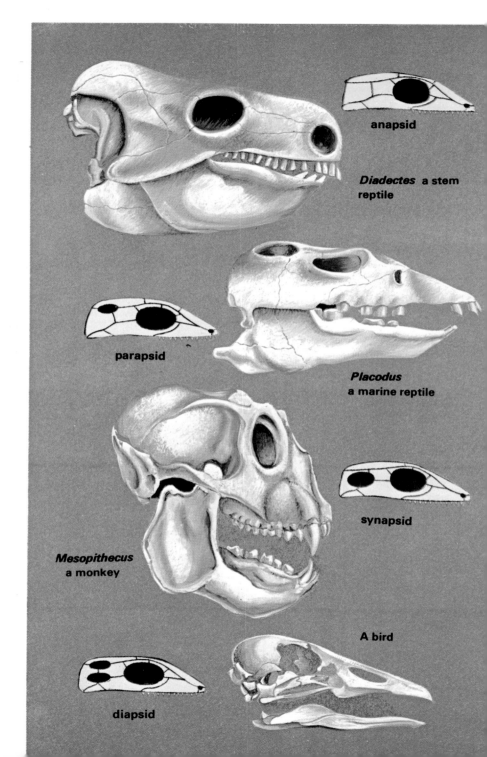

anapsid

Diadectes a stem reptile

parapsid

Placodus a marine reptile

synapsid

Mesopithecus a monkey

A bird

diapsid

Changes in reptilian skull structure

One very important feature of the reptiles is the evolution of the temporal region of their skulls. The structure of this region has given rise to the identification of four main groups and by close investigation linking relationships have been made between the groups. Even so, although the temporal openings illustrate certain evolutionary trends, some zoologists think such a classification is misleading.

The most primitive reptiles, the cotylosaurs, had a complete bony roofing in the temporal region—the *anapsid* skull. The turtles *(Chelonia)* have a similar arrangement, although the bones at the back of the skull are reduced.

From this anapsid condition probably evolved two principal types. The *synapsid* group had a single opening (fossa) in each side bounded by the post-orbital and squamosal bones above, and the squamosal and jugal bones below. The reptiles with this type of skull were on the evolutionary line which led to the mammals. The skull was modified by a reduction in the bones and their positioning.

The other modification of the primitive anapsid condition gave the *diapsid* skull. Here, there are two openings in the temporal region. This type is found in many reptiles, including the crocodiles, the dinosaurs and the pterosaurs. A modification is found in the lizards and snakes. It is from the diapsid skull that the bird skull was modified. The birds lost the post-orbital bone so that the two openings in the temporal region and the eye form one large opening.

In other groups only a single opening and arch is present, similar to the synapsid condition. However, in these cases the opening is situated high on the skull, surrounded by the parietal bone above and squamosal and post-orbital below. This type of skull is termed *parapsid,* sometimes *euryapsid,* and is seen in the ichthyosaurs and plesiosaurs.

Although the classification of reptiles by their skulls is in some ways artificial, it does serve to indicate the main lines of their evolution and how the birds and mammals are descended from certain types.

Diagrammatic fossil reptile skulls in blue reveal patterns of evolution found throughout the vertebrate world.

Archaeopteryx

The origin of flight

The air was the last major environment to be colonized by vertebrates, and in this conquest the reptiles played an important part. For some 100 million years the insects had held the monopoly in the air, but beginning with Jurassic there are traces of flying reptiles. The pterosaurs or winged lizards showed the three essentials necessary for life in the air—their skeleton was strong but light, they had wings and their sight was good because of the large cerebral hemispheres which control sight and muscle coordination. The Jurassic species were small with toothed jaws and long tails, but by Cretaceous times they often had a 25-foot wing span and long, pointed toothless beaks, often longer than the vertebral column. They seem to have been gliders and soarers rather than active fliers, as the wings were less robust than in birds and bats.

The pterodactyls are more common in the Jurassic than in the Cretaceous. They seem to have paralleled the evolution of the birds from a thecodont stock, but there is no evidence that they were feathered, although the wing had a membrane.

The second group was the evolution of the warm-blooded birds from the diapsid archosaurs. There is no detailed evidence of the transformation of cold-blooded reptiles into warm-blooded birds. However, certain fossils do give ideas of the intermediate stages. The oldest-known birds are *Archaeopteryx* and *Archaeornis* found in the upper Jurassic of Bavaria. Although these early birds are fully feathered, in many ways they still resemble reptiles.

During this time, birds underwent great changes and the fossils of the Cretaceous showed them to be true birds. A flightless swimmer, *Hesperornis* was a wingless, streamlined bird with a long beak and webbed feet. It probably plunged beneath the surface in search of prey. Another species, *Ichthyornis,* was a powerful flier, probably feeding on the shoreline or swooping down to pick up fish at the surface.

The origin of flight is still open to much discussion. One line of thought is that the early birds were tree dwellers, and the developing wings broke falls. These leaps subsequently became glides. Another theory is that the ancestors were terrestrial, and feathered wings and tail developed to increase speed over the ground until they eventually lifted into full, flapping flight.

The diving *Hesperornis* of the Cretaceous

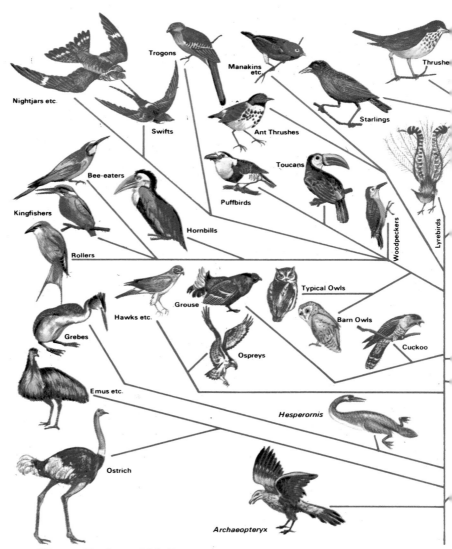

The radiation of birds

The basic bird structure has become modified to produce the great variety of modern birds. Due to the few fossil remains, the direction of change has been built up from the study of present-day forms. The radiation took place in the Cenozoic. The widely differing habitats of the birds range from the powerful birds of prey, the penguins, the waders of the shore

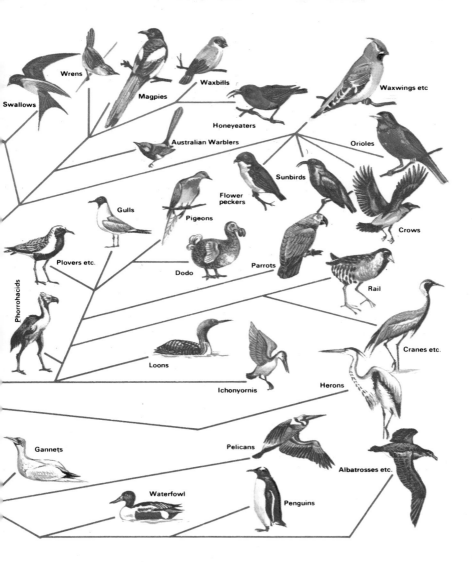

and estuary, to the game birds. The smallest bird is the Cuban bee hummingbird, weighing 0.07 ounces; the ostrich is the largest and may weigh 300 pounds. The largest now-extinct birds were the elephant birds of Madagascar and the moas of New Zealand. The vast number of genera are grouped into over 40 orders, and the Passerines or perching birds comprise about half of all the species.

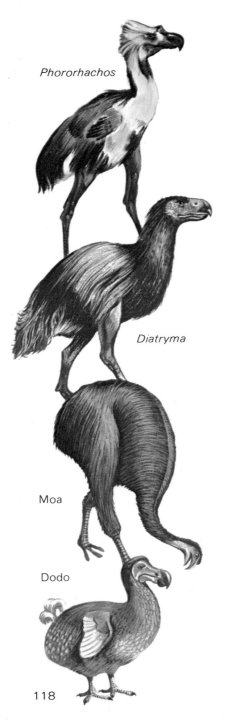

Phororhachos

Diatryma

Moa

Dodo

Extinct flightless birds

The largest extinct birds were all flightless. At first it was thought that they were all related, but this idea is no longer believed. As they evolved terrestrial habits, their increase in size and ferocious nature protected them from their enemies.

Diatryma was a formidable carnivorous ground creature of the early Eocene of North America and Europe, some 7 feet high with reduced wings and a massive head.

The South American *Phororhachos* resembled *Diatryma* closely. Some of their heads were horse-sized, and fossils occur in the Oligocene, Miocene and Pliocene rocks in Argentina.

The first bone of a moa was brought to London in 1838 from New Zealand. Evidence shows that natives killed moas for food and cooked them in earthen pits 5 feet in diameter. The largest birds were 12 feet tall.

The dodos were abundant for some time on Mauritius. The numbers of these swan-sized birds, related to pigeons, were reduced by Portuguese explorers in 1505 and later by the Dutch in 1598. By 1691 the dodos were extinct.

Living flightless birds

Ostriches, rheas, emus, cassowaries and kiwis are collectively known as the ratites.

Today ostriches are found in Africa and Arabia, although fossil evidence shows that their ancestors were present in the Eocene in Switzerland. Their well-padded, two-toed feet enable them to run quite rapidly over stony and sandy deserts.

The rheas of Argentina are sometimes called ostriches, but they are not closely related. They have been found in South American Pliocene formations.

The emu and cassowary group is restricted to Australia and New Guinea. Their wings are even more reduced than those of the ostrich and rhea.

Relationships of the penguins of the southern hemisphere to other birds are not clearly understood. It seems they may have had a common ancestry with the oceanic albatrosses, shearwaters and fulmars, possibly in the late Cretaceous or early Cenozoic. Even the earliest forms seemed incapable of flight. Opinions differ as to how they evolved to become so beautifully adapted for an aquatic life.

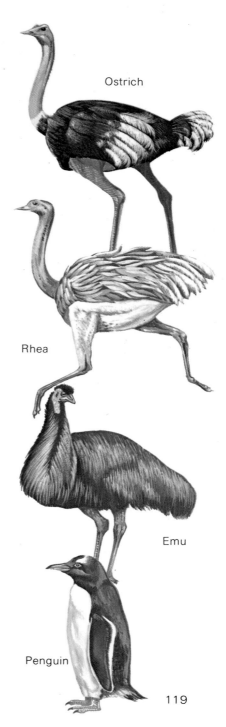

Ostrich

Rhea

Emu

Penguin

119

The origin of mammals

The end of the Cretaceous marks the turning point for the reptiles, as the mammals began to replace them as the dominant vertebrates. There seems little doubt that the chief factors were the climatic and geographical changes which began in the early Tertiary. Temperate conditions replaced the previously widespread warm conditions. The cold-blooded reptiles lacked good insulation; they had become adapted to the warmth and were unable to adjust to the new conditions.

Climatic changes also affected the flora. The cycads had been the typical plants of the Mesozoic, but these only survived in the warmer parts of the world. They had provided the bulk of the herbivorous reptiles' food and, as food became

Ophiacodon was one of the earliest mammal-like reptiles.

scarce, a decline in the reptiles followed. The small size of the cranium may have also been a factor, or muscular movement and coordination may have become difficult with increasing size.

Mammals had been slowly evolving during the dinosaur dominance. Reptiles with mammalian features existed from early Triassic times. A typical example is *Cynognathus,* some 7 feet long. Outstanding mammalian features were differentiated teeth and limbs that held the body in a mammalian posture. It was not until very late in the Triassic, however, that true mammals evolved.

The surviving mammals which, like reptiles, lay eggs, are the

Echidna

Australian duckbilled platypus (*Ornithorhynchus anatinus*) and the spiny anteater or echidna (*Tachyglossus aculeatus*). They are so different from other mammals that they must have left the main mammalian stock in the early Mesozoic. Their organization may show us many of the characteristics of the Mesozoic mammals. Today they are highly specialized creatures; both are toothless as adults, although the young platypus has a few tooth rudiments. The absence of teeth is compensated for by the development of a broad, horny bill. The echidna is protected by stout spines. It has powerful digging limbs and a long snout. Its mammalian characteristics include fur and it nurses its young even though it lays eggs reptilian style.

Platypus

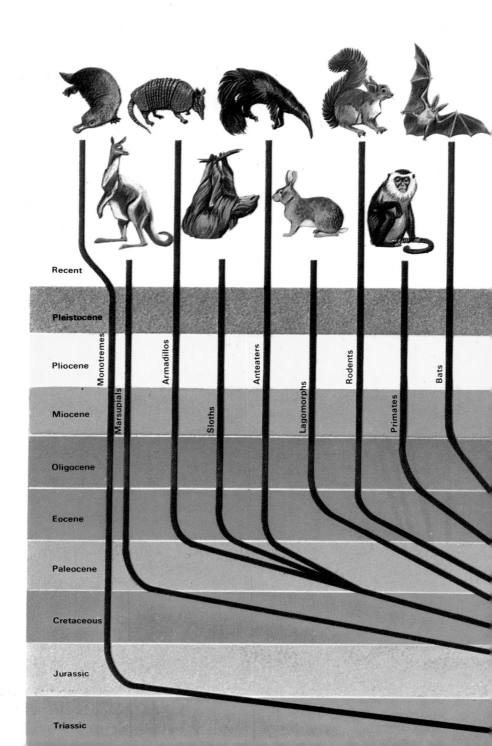

Recent

Pleistocene

Pliocene

Miocene

Oligocene

Eocene

Paleocene

Cretaceous

Jurassic

Triassic

Monotremes

Marsupials

Armadillos

Sloths

Anteaters

Lagomorphs

Rodents

Primates

Bats

Insectivores

Cetaceans

Artiodactyls

Carnivores

Horses

Chalicotheres

Elephants

Condylarths

Titanotheres

Perissodactyls

Uintatheres

Amblypods

Tricondonts

Symmetrodonts

Multituberculates

Therapsids

Evolution of mammals

Bandicoot   American opossum

Pouched mammals — the marsupials

The pouched mammals show essential similar characteristics to placental mammals but are primitive in many other
characteristics and undoubtedly diverged from the mammalian stocks at some early age. In some ways they portray what
mammals may have been like·in the late Cretaceous.

There are some 230 living species of marsupials, all varied
in habit and design, paralleling in the isolation of Australasia
the adaptive radiation accomplished in other parts of the world
by the placentals. Today, the majority of marsupials are found
in Australia, with a few representatives in North and South
America, but in the Eocene they definitely occurred in Europe
and presumably became restricted by the competition imposed by rapidly evolving placentals.

The South American continent is now connected to the rest
of America by the narrow·Isthmus of Panama. However,
during the early Tertiary it was unconnected and the link was
not re-established until quite late in the Tertiary. Except for

Tasmanian Wolf

Kangaroo

Koala

Phalanger

caenolestids, rat-like marsupials, all the marsupials found today in South America are opossums and some 72 species are recognized. Fossil evidence shows that during the isolation, the marsupials radiated in several directions. Some paralleled the carnivorous cats and dogs of other continents. When North American connections became re-established, placentals invaded, quickly causing the extinction of the carnivorous marsupials.

Australian marsupials have survived because the large continent was isolated from the rest of the world in the Cretaceous and remains isolated today. Pouched mammals had entered before the isolation and it was only through man's agency that other land mammals entered (excluding bats and rats which apparently originated from the East Indies). Thus a curious and fascinating fauna was able to evolve and, again, many forms have paralleled groups of higher mammals evolved in other areas of the world.

Marsupials successfully radiated into a multitude of forms in habitats free from competition from placental mammals.

Tasmanian Devil

Wombat

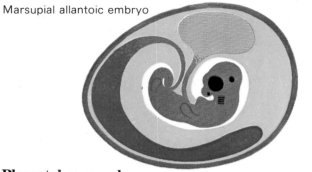

Marsupial allantoic embryo

Placental mammals

The evolution of primitive placentals marks the final stage in the general line of mammalian ascent, the placenta acting as an intermediate structure by which the developing embryo is nourished. The placental method of reproduction results in the birth of young at a much more advanced stage than in marsupials. The factor of longer prenatal development and the period of post-natal care and training which follows, together with the much larger and efficient brains which most groups possessed, are features of great significance in the dominance of the placental mammals.

The earliest placentals appeared in the Cretaceous, about the same time as the marsupials, which seems to point to a common ancestor. They were insectivores, the forerunners of the living moles, shrews and hedgehogs. Some show similar

The placenta enables the embryo mammal to develop to a more advanced state before birth.

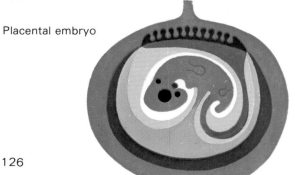

Placental embryo

126

Deltatheridium, a reconstruction of a fossil form

structures and appearance, whereas others were definitely more primitive in form. All were small, mostly unobtrusive, nocturnal in habit and totally insignificant in comparison with the dinosaurian giants among which they lived, and with the mammals which later evolved from these primitive placental stock beginnings.

Deltatheridium is a fossil insectivore found in the upper Cretaceous of Mongolia and shows characteristics very close to those of the ancestor not only of insectivores but of all placental mammals. Its skull was tubular and elongated, but it lacked the specialized snout of some of the modern forms.

From these early mammals various forms radiated, adapting and filling every suitable environment. Many of the mammals became larger. The limbs became longer and specialized for various modes of locomotion. The teeth number was reduced in many and their form became modified to utilize most effectively a particular diet.

Modern insectivorous shrews resemble their remote ancestors.

Flesh-eating mammals

The earliest Cretaceous mammals were most probably insectivorous, and it is therefore not surprising that some of the descendants became flesh-eating. Indeed, practically all modern flesh-eating mammals since have evolved from this single stock.

The major changes in mammals of carnivorous habits were in the design of the teeth. The carnivore kill is made using mainly the teeth and these have to pierce stout hide, cut tough tendons and grind hard bone. The fact that flesh is quite easily digested means that it need not be well-chewed. Thus it is the front part of the mouth that is highly adapted. The front incisors are the biting and tearing teeth; the side canines are long and pointed, piercing and stabbing weapons in all the carnivores. The cheek teeth are usually reduced in number and those that are present have sharp ridges and pointed cusps rather than flat surfaces. In many flesh eaters, such as dogs and cats, a very specialized pair of teeth called 'carnassials' have been developed on either side of the jaw. The teeth do not meet directly but slide past each other, scissor-like, slicing or cracking very tough tissue.

The skeleton has a more primitive basis, for it must remain

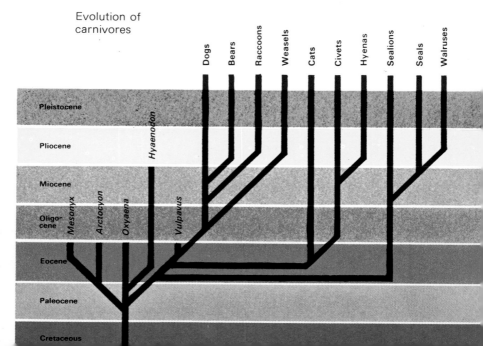

Evolution of carnivores

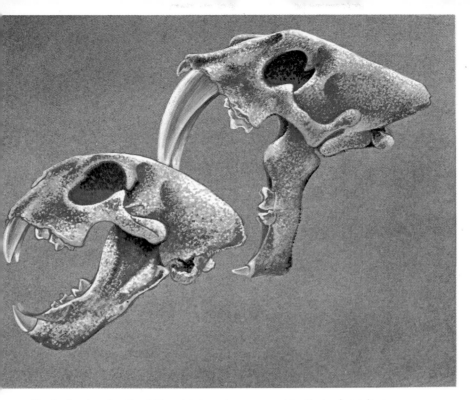

Skull of saber-toothed 'tiger' (*above*) compared to that of modern cat (*left*) to show modifications for striking and biting

supple to enable the animal to move fast. It does not require a reduced toe number or hooves, as the claws are important in climbing trees and for attacking and holding prey. In some forms the 'thumb' and big toes are lost.

The flesh-eating forms were dominant during the first epochs of the Tertiary. These are referred to as archaic carnivores or ceodonts and are now all extinct. The range in size was great, some compared with modern weasels while others became very large. The adaptive radiation which took place in this hunting order is wide. For example, the seals, sea lions and walruses are marine carnivores that have existed since the Miocene. They are thought to be descended from ancestors within the weasel family. The bears were a Miocene offshoot from the dog stock. The most modified are the civets and mongooses (Viverridae), the hyenas and the true cats (Felidae).

A chalicothere

Hoofed mammals — odd-toed ungulates

During the Paleocene a number of animals abandoned the insectivorous habit and began to eat plants. The early ungulates are known as the condylarths and resemble the early carnivores, with a long body and limbs that are short and primitive in structure. However, the cheek teeth evolved for chewing vegetable food and, although all the toes were present, each was capped by a small hoof. From these early herbivores two groups of ungulates developed — perissodactyls and artiodactyls.

Living odd-toed ungulates, perissodactyls, include the horses, rhinoceroses and the tapirs. The foot symmetry has the axis through the middle toe, and there has been a tendency for reduction to a single toe.

Of the early ungulate forms, the horse-like titanotheres began in the Eocene and rapidly evolved to gigantic sizes. For unknown reasons they became extinct by the Oligocene.

Tapir

Rhinoceros

Not uncommon in the Tertiary were the chalicotheres. They had teeth similar to those of titanothere, and they were suitable only for soft plant food. Their toes ended in huge claws.

The only living representative of the tapirs are confined to the tropics of South America and the Malay region. In the Tertiary they were widespread in North America and Europe, disappearing from these regions in the Pleistocene.

Today rhinoceroses are confined to the Old World tropics. The early forms of the Eocene were slim, small, horse-like in build and were without horns.

A titanothere

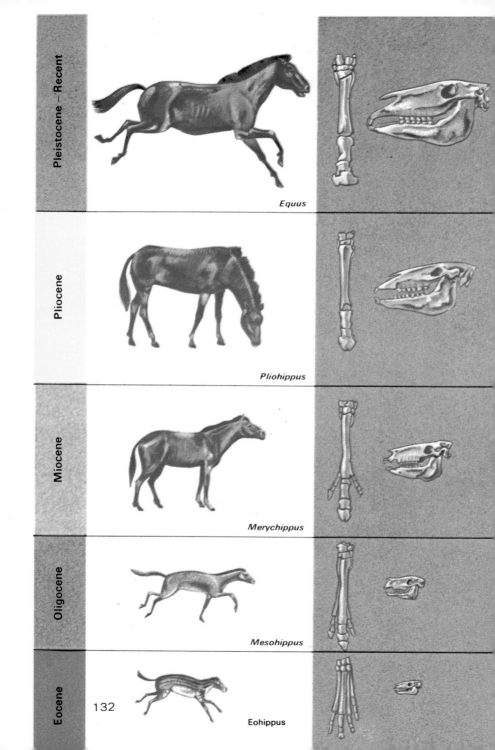

Pleistocene – Recent

Equus

Pliocene

Pliohippus

Miocene

Merychippus

Oligocene

Mesohippus

Eocene

132

Eohippus

Evolution of the horse

Probably the most widely known of all evolutionary sequences is that of the horse. Tertiary deposits from North America where most of the horse evolution took place, contain a very good fossil series. The earliest horses were the size of a small dog and are known from the earliest Eocene. From the chart illustrating their subsequent evolutionary trends we can observe four main changes. These involve an overall increase in size, lengthening of the limbs and modification of foot structure, changes in the skull and teeth, and an increase in both the relative size and complexity of the brain.

Whereas the earliest horse, *Hyracotherium*, commonly known as Eohippus, stood 12 inches at the shoulder; a modern workhorse may stand well over 5 feet. Although there are obvious advantages in an increase in size, there are also several problems. The great increase in weight involves increased stresses on the limbs which must therefore evolve strengthened bones. Also, feeding habits become more complex as larger amounts of food are required. The modern horse is by no means just an enlarged edition of Eohippus. The whole animal has undergone drastic changes in various structures, on the whole related to the overall increase in size.

The teeth changed in design and size due to the increased bulk and the change in feeding habit. They gradually increased in length, surface area and structural efficiency. In the Miocene they show radical changes, probably due to the horse becoming a grazer rather than a browser in feeding habits. This was related to the change in environment at this time. There is abundant fossil grass seed evidence that during this period there was a transition from hardwood forests to open forests and prairie.

With the change in habitat from forest to prairie, there was a reduction in toes, from a 'pad-footed' type to a 'spring-footed' type. The reduction to a single hoof and the lengthening of the limbs also reflect the change to life on the plains. The single hoof would have been highly unsuitable for a forest environment but was ideal for high running speeds on the firmer ground of the prairies.

Development of limbs, skulls and form during evolution of the horse. Note the reduction in the number of toes which allowed greater speed over hard ground. The reconstructions are to scale.

133

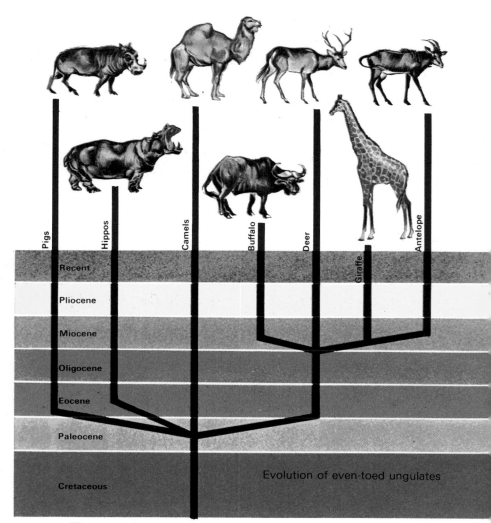

Evolution of even-toed ungulates

Even-toed ungulates

Even-toed ungulates, or artiodactyls, have for many thousands of years been of extreme importance to man and to the carnivores whose lives depended on them. Although present at the beginning of the Eocene, it was not until later that they became a prominent and diversified group. There seems to have been a distinct division into pigs, camels and the true ruminants (deer and bovids).

Pigs and their relatives are, in many ways, the most primitive of living even-toed ungulates. They are four-toed animals, although the side toes are reduced. The limbs have not elongated to the extent of the limbs of other artiodactyls.

The hippopotamus is distantly related to the swine found in the Pleistocene in regions as far removed as England and China. Only two species survive today, both in Africa.

Camels and llamas were an early group to evolve and have had a long distinct history. North America was their original home, however, after the true camel had become established in Asia, and the llamas in South America, the camel died out in North America.

The success of the ruminants is undoubtedly largely due to the four-chambered stomach which allows the animal to digest tough grasses. The complex arrangement allows ruminants to gulp large amounts of food quickly, retire to a safe spot and regurgitate it from the first two chambers, to be chewed and digested at leisure. The ruminants were the last important mammalian group to evolve, most of them branching off in the Eocene. The antlered deer, giraffe and okapi, and the bovids (cattle, sheep and goats) have even more complex stomachs.

One might question the success of the even- over the odd-toed ungulates. The teeth, the brains and the feet played an important part, but probably the development of the stomach was the true reason for the marvellous success of the artiodactyls.

Toe variation among even-toed ungulates

| Hippopotamus | Deer | Camel |

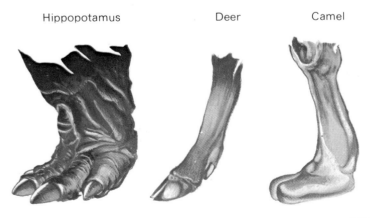

Elephants and relatives

Fossil evidence shows that in the past the elephants have been adapted to all varying types of climate.

The earliest form traced is that of *Moeritherium*, a small creature showing few of the elephantine characteristics we imagine today. There was no trunk (perhaps a pig-like snout), and all teeth were present, though signs of an enlarged upper and lower pair were visible. Certain authorities debate whether or not it should belong to the elephant group as it had these primitive characteristics. Gradually an increase in size took place; the jaws shortened, tusks evolved—the upper ones becoming free and curved, the lower ones eventually reducing—and a highly flexible trunk formed. The two modern elephants and the mammoths derived from certain mastodonts of Miocene stock. The evolutionary pattern is quite complex, and the sequences are mostly guesswork.

The early mammoths originated in Asia and soon spread over Europe. Descendants migrated to North America where they gave rise to gigantic forms. The woolly mammoth (*Mammuthus primigenius*) was adapted to open country, and was quite common during the last glacial period. Its true appearance is well known as a result of discoveries of whole specimens found in the ice of Siberia.

Although there are numerous other differences between the modern African elephant (*Loxodonta africana*) and the Indian elephant (*Elephas maximus*), the usual identification is that the African has large ears and the Indian has small ears.

The hyraxes (Procaviidae) are thought to be relatives of the elephant, having branched from the very early elephant lineage probably in mid-Paleocene times. Superficially they resemble rabbits, but internally there are certain distinguishable links with the elephant.

The sea cows (Sirenia) of today include the dugong of the Indian Ocean and the Red Sea and the manatee of the shores of the tropical Atlantic. They are aquatic and browse on coastal vegetation. In the Tertiary they were quite widespread. Resemblance to older fossils unmistakably link the sirenians to the the elephants.

Evolution of the elephant

Elephant

Mastodon

Mammoth

Dinotherium

Trilophodon

Moeritherium

137

Rat

The diversity of mammals

Rodents are the most numerous of all mammals and have migrated during their evolution to live in nearly every corner of the earth, from the Arctic to the tropics. Although they have a small brain, they have proved to be the most successful of mammals because of their adaptability and extremely rapid breeding.

The whales are almost completely adapted to an aquatic life. Although their ancestry is uncertain, they are thought to have diverged from early placental eutherian stock. Fish-like in shape, their internal adaption shows the most extreme specializations of any mammal. Vestigial hind limbs, not connected to the spine, of ancestral terrestrial forms can be found in whales. The forelimbs are flipper-like.

Although there is no intermediate fossil evidence, bats are

Sperm whale

Horseshoe Bat

thought to have branched from a primitive stock of the shrew-like insectivores. There existed several different species in the Eocene, and today there are some 750 species.

The armadillos, anteaters and sloths, collectively known as the edentates, are peculiar to South America. The anteaters feed on the termites of tropical regions, using strong claws to rip nests apart.

The sloths hold on to the branches of trees, using their long curved claws, hanging in the characteristic upside-down position. Tree-sloth fossils are very rare, but whole skeletons have been found of the extinct giant sloth, *Megatherium*. This sloth of the Pleistocene period grew to elephantine proportions.

The armadillos are protected by bony plates covered with horn. The giant armadillo, *Glyptodon* of the Miocene, shows that some grew to over 10 feet long.

Sloth

139

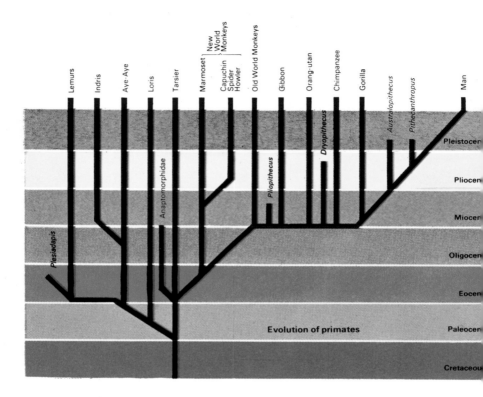

Lemurs · Indris · Aye-Aye · Loris · Tarsier · Marmoset · New World Monkeys · Capuchin Spider Howler · Old World Monkeys · Gibbon · Orang-utan · Chimpanzee · Gorilla · Australopithecus · Pithecanthropus · Man

Anaptomorphidae · Pliopithecus · Dryopithecus · Plesiadapis

Pleistocen · Pliocen · Miocen · Oligocen · Eocen · Paleocen · Cretaceou

Evolution of primates

Primates

The order Primates includes apes, monkeys, lemurs, bush-babies, tarsiers and tree shrews. Fossil evidence to link the families is scarce, due to the fact that most of the members are arboreal. The limbs of primates are quite primitive when compared to other mammalian specializations. In most lower forms the primitive long tail is retained as a balancing organ, but in some higher forms it becomes reduced or lost entirely. A tree-dwelling life demanded good vision and muscular and nerve coordination. The brain centers for these activities became greatly enlarged compared with those of their ancestors, while the sense of smell degenerated.

The structure of the tree shrews shows them to be inter-mediates between insectivores and primates, but their brain is much more advanced and complex than that of the insectivores. The tree shrews have changed little since Paleocene times when they first branched from the insectivores.

From the tree shrews arose the more advanced lemurs and tarsiers. All are arboreal, feeding on fruit and insects, and now they survive mainly in restricted areas. The lemurs are found only in Madagascar, the lorises in Asia and the galago or bush-babies and potto inhabit areas of Africa. In early times they were all widespread, but they became restricted due to competition with the evolving monkeys. The curious aye-aye has claws on all toes except the big toe and procumbant incisors like a rodent; these specializations allow it to obtain bark-dwelling insects. The lorises of India move in a slow, deliberate manner and are unable to jump, as do the pottos of Africa, but the galagos with elongated back legs are able to jump magnificently.

The one living form of the tarsiers, *Tarsius,* is of the East Indies and Philippines. Fossil forms date from the Paleocene, and 25 genera have been identified from North American and European deposits. Too specialized to be on the direct line of descent of the higher primates, *Tarsius* does, however, show links. The neck is very mobile as the large, forward-directing, nocturnal eyes move very little. The ankle bones of the foot are greatly elongated thereby lengthening the foot and giving *Tarsius* powerful, long hind legs, for jumping.

Nocturnal tree-climbers, the aye-aye (*top*) and potto (*bottom*). The potto is the African equivalent of the loris.

Mandrill   Patas

Old world monkeys

These are the monkeys of Asia and Africa, evidence for which dates back to the Oligocene. The more primitive forms are arboreal, but the baboons and their relatives show a tendency toward a ground-dwelling life. Two groups are distinguished today; one group has cheek pouches and the other has a complicated type of stomach (similar to ruminants).

The cheek-pouched groups are frequently found in zoos. The macaques, rhesus monkey and their relatives are specialized, with less arboreal habits, longer grinding teeth and somewhat longer faces. The baboons, mandrills and drills have a characteristic dog-like snout due to the elongated teeth evolved for a herbivorous diet.

The langurs of Asia are examples of the second group. One member is the proboscis monkey with its wonderfully elongated pendant nose.

Gelada   Rhesus

Marmoset

Uakari

New world monkeys

The South American monkeys are known as flat-nosed monkeys because their nostrils are widely separated and face more sideways than downward and forward.

The smallest forms are the colorful marmosets, squirrel-like, with thick long fur, bushy tails and claws. The marmosets and tamarins, unlike most other anthropoids, give birth to two or three young at a time.

The cebids are the other South American monkey group. The spider monkey is a typical member with a prehensile tail. Capuchin monkeys will eat fruit and insects, whereas others feed only on fruit. The uakaris are bizarre cebids with bald heads and long brown hair.

In all monkeys the eyes are large and facing upward, giving binocular vision. An increased brain size allows for a high level of learning and intelligence.

Spider monkey

Capuchin

143

Apes

Today the only ape representatives are the gibbons, orang-utan, chimpanzee and gorilla. All are more advanced than monkeys, notably in the larger and greater complexity of the brain and the different tooth pattern. Because of their large size apes can no longer run along the tops of branches in monkey fashion. They have become great movers by swinging with their arms which are longer than their legs and are provided with very powerful muscles. These changes have meant changes in the body structure; the chest has widened, the neck and limbs lengthened, and the head enlarged.

There are close anatomical and physiological resemblances between the apes and modern man. There is no structure in the gorilla that cannot be found in man. The relative development and size of certain body parts are quantitatively different, the most important ones associated with locomotion and brain size. This does not mean that man evolved from the apes. The two groups probably diverged from a primitive basic stock in the early Miocene.

Gibbon (*top*) and
Chimpanzee (*bottom*)

Fossil evidence of the apes is not common, but examples show that these primates were widely distributed throughout Europe, Asia and Africa during the middle and latter portions of the Cenozoic.

Parapithecus is described by some authorities as an early Old World monkey, whereas others regard it as an early ape. Perhaps it was common to both; one cannot, at present, state precisely where it should be placed. In the same Egyptian beds of the lower Oligocene age, a jaw was identified as that of a small primitive ape, *Propliopithecus*.

The earliest apes to appear after *Propliopithecus* were *Limnopithecus* and *Proconsul* from the lower-Miocene sediments of Africa. These apes, called dryopithecines, are believed to have included ancestral forms that eventually led to man.

There seems to have been apes living in Asia during the Pleistocene epoch that were larger than the gorillas. Today the gorilla and chimpanzee inhabit Central Africa, the orang-utan lives in Borneo and Sumatra, and the gibbons inhabit the tropical forests of Indonesia and southeast Asia.

Orang-utan (*top*) and Gorilla (*bottom*)

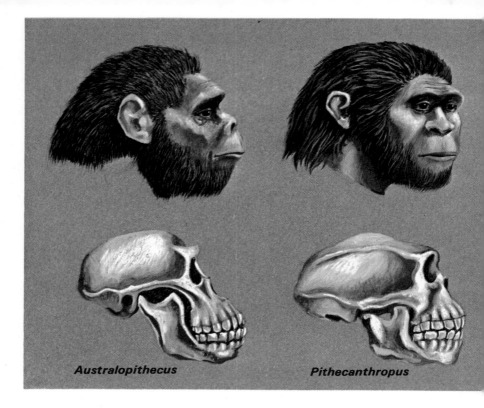

Australopithecus **Pithecanthropus**

From apes to man

The fossil record of man is incomplete, but a general outline is beginning to emerge. From the end of the Pliocene into the Pleistocene (from 2 million to 700,000 years ago), an ape-like homonid lived. The first fossil was found at Taung in Bechu-analand in 1924. Since then, a number of specimens have been found in East Africa. They vary somewhat, but all belong to the same group, the southern apemen, or australopithecines. The name *Homo habilis* is given to some members of this group by some authors. The australopithecines had brains that were one-third the size of modern man. They probably did not use fire or build shelters, but they did walk erect and made crude tools.

From about 600,000 years ago to the appearance of modern man, a new homonid skeleton appeared. Pithecanthropus, or *Homo erectus,* had a skeleton that was very much like that of modern man but with a massive skull, large teeth and a smaller

146

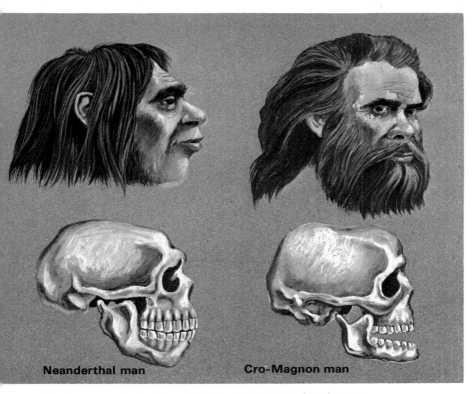

Neanderthal man Cro-Magnon man

Skull development and probable appearances of early men

brain. The brain of Pithecanthropus was about twice the size of that of *Australopithecus*. Pithecanthropus has been found with stone tools and charred pieces of wood, indicating that he made tools and used fire.

Near the end of the last Ice Age, a distinctive type of man appeared. Known as Neanderthal man, he possessed very large brow ridges and, in general, a massive skeleton. Neanderthal man had a definite ape-like appearance and was first thought to be a primitive stage in man's development. He is now generally considered to have been an aberrant sideline in the evolution of man. About 40,000 years ago the neanderthals were replaced by a thoroughly modern human type, *Homo sapiens*.

Because of man's abilities to think and to reason logically, he has been able to migrate to fill practically every corner of the earth. He now can control his own evolution and more or less decide the future of all other forms of life.

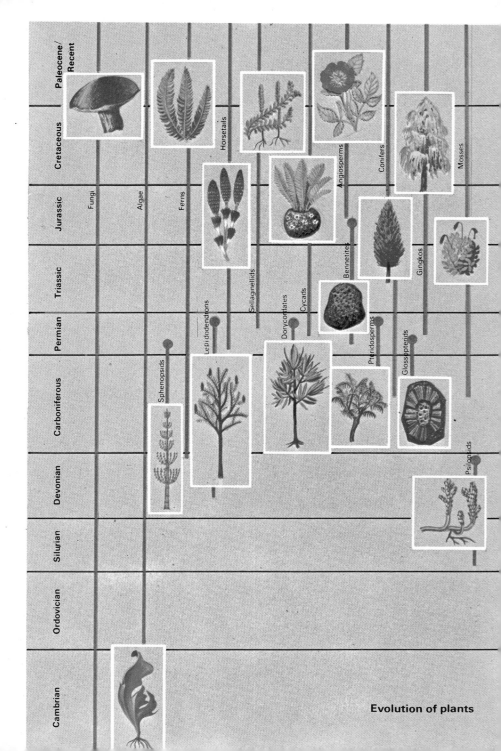

Evolution of plants

Paleocene/Recent

Cretaceous

Jurassic

Triassic

Permian

Carboniferous

Devonian

Silurian

Ordovician

Cambrian

Fungi

Algae

Ferns

Horsetails

Selaginellids

Lepidodendrons

Sphenopsids

Dorycordiates

Cycads

Bennerties

Pteridosperms

Angiosperms

Conifers

Gingkos

Mosses

Glossopterids

Psilopsids

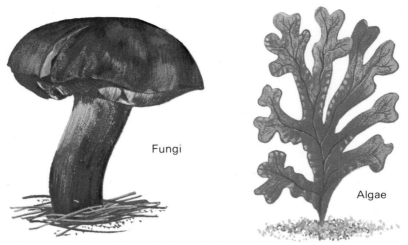

Representatives of the primitive plant group Thallophyta

Evolution of plant life

As plant and animal life evolved in water, the properties of this medium imposed certain restrictions on the evolutionary possibilities of organisms living there. The first plants were unicellular or multicellular organisms with simple structure, without differentiation into stems, leaves and roots. They are known as the Thallophyta and are represented by algae of various kinds, including seaweeds and stoneworts.

Plants evolved no further until they colonized the land. There is evidence in the Cambrian period of spores from land plants, but in the Devonian the land was already partially covered by thallophyte plants. At first the thallophytes lived in shallow water. So that the plants could stand erect, resist desiccation, take in atmospheric carbon dioxide and oxygen, and obtain food materials, the thallophytes evolved a stiff outer layer perforated with stomata. The roots absorbed water and the mineral salts dissolved in it from the soil. The method of reproduction also had to adapt. In water, the thallophytes reproduce sexually by eggs and sperm or asexually by spores, and in order to reach the egg and fertilize it, the sperm must swim through the water. On land, therefore, a thin film of moisture is required to ensure fertilization.

The Psilophyta were the simplest land plants known–well-preserved remains have been found in the Devonian period.

149

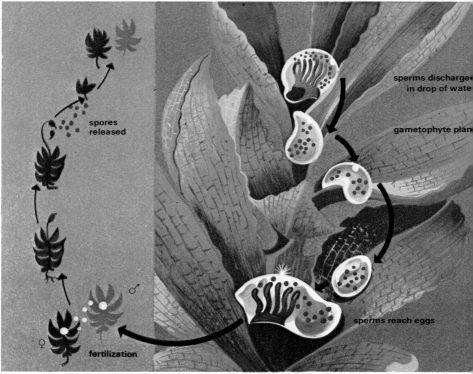

Labels within image: sperms discharged in drop of water, gametophyte plant, sperms reach eggs, spores released, fertilization, ♂, ♀

Alternation of generations in a bryophyte moss

One, *Rhynia major,* consisted of a stem continuous with an underground stem or rhizome which was not differentiated into a root but bore absorbent root-hairs. The stem branched and at the end of some branches were spores in groups of four, as in ferns. Stomata, small openings that allow the flow of air, are visible in the stems, just as in all land plants. Rudimentary leaves were found in a related plant, *Asteroxylon mackiei.* These Devonian plants evolved from the marine algae. Evidence to establish that the Psilophyta are directly ancestral to any higher land plants is lacking. However, they do show the appearance of the plants that first colonized dry land. They could live only in damp places because a film of moisture is necessary to enable the sperm to pass to the egg. In the classification of plants the method of reproduction usually positions the individual species.

The **Bryophyta** (mosses and liverworts) have a fairly simple cellular structure which restricts their size. They have not evolved very far and have developed a reproductive method known as alternation of generations. The moss plant is known as the *gametophyte* as it produces the eggs and sperm. The fusion of these give rise to the small *sporophyte* which lives parasitically on the moss leaf. It grows and liberates spores which germinate into moss plants. As the gametophyte is haploid, it is less well provided with possible variants in the genetic make-up.

In the more advanced **Pteridophyta** (ferns, horsetails and selaginellids) the plant is the sporophyte, while the gametophyte is minute. The sporophyte has the diploid condition of chromosomes and so is well equipped to provide variants. Thus natural selection has been able to work, and the results are evident when we look at the plants themselves. They have evolved 'vessels' running through the root and stem systems for transport of water and food substances and are thus sometimes called the 'vascular cryprograms'. The life-cycle diagram illustrates the alternation of generations in a fern, but the sequence is the same for all pteridophytes.

Alternation of generations in a pteridophyte fern

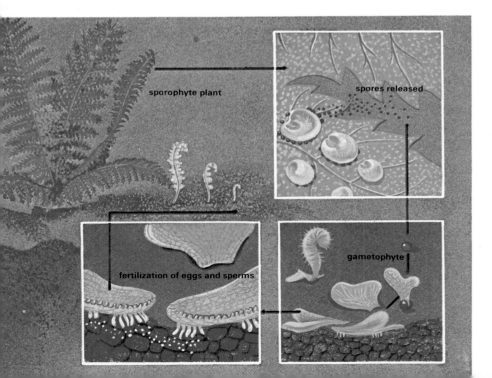

sporophyte plant

spores released

fertilization of eggs and sperms

gametophyte

Cryptograms and phanerograms

The thallophytes, psilophytes, bryophytes and pteridophytes are grouped together and called **cryptograms** ('hidden marriage') as they do not produce by visible seeds, but by spores. Selaginellids, such as *Lycopodium,* show a transition stage in the evolution of seeds. They have two kinds of spores; large 'megaspores' develop into female prothalli bearing egg-producing **archegonia**, and small 'microspores' which develop into male sperm-producing **antheridia**.

Certain cryptograms evolved to produce types which suppress the gametophyte generation completely. It does not exist as a separate structure, but is represented by a few cells and nuclei only when the spores are produced. The pollen grain is the microspore and one of its nuclei develops to be the sperm. The egg-sac is the megaspore with the egg inside it.

The pollen grains are shed into the air and usually transported by the wind. The megaspore is not shed but is retained within its spore-case, or ovule. The ovule is exposed to the air and at a certain part of its coat allows a pollen grain to come into contact with the megaspore. The pollen grain is able to penetrate the egg by growing a pollen tube and a nucleus, acting as the sperm, passes down as it grows until the egg is reached. No film of water or moisture is required, and it is this advancement in the reproductive mechanism that enabled plants to spread and colonize dry land.

Once fertilized, the egg becomes an embryo and develops within the ovule which forms the seed. If conditions and environment are suitable, the seed's embryo will germinate and produce a seedling and eventually a new plant. In this method of reproduction, seeds are formed and eventually expelled, and the plants are known as **phanerograms** ('conspicuous marriage').

'Gymnosperms' (naked seeds) are the most primitive seed-bearing plants. Among the living members are the cycads, the ginkgo, the conifers and the pteridosperms and Cordaitae. True fossilized gymnospermous wood in mid-Devonian rocks has a structure as highly organized as the secondary wood of living conifers. Several groups seem to have evolved during the Triassic and Jurassic to maximum development and then disappeared in succession. Jurassic fossils show relatable characters to the living *Araucaria* (monkey-puzzle tree).

The gymnosperms bear male and female cones. The female cones are formed from the grouping of leaves bearing the ovules, and the male cones from the grouping of the stamens.

Flowering plants

About a quarter of a million living species are found in the group of flowering plants. Here evolved the complete protection of the ovule by its enclosure inside an 'ovary', a chamber formed from the carpellary leaves that bear the ovules. From the ovary, projecting upward, is a thin cylindrical style. The end, known as the stigma, is sticky so that pollen grains touching it will adhere. The pollen grains grow down the stigma shaft and the first to reach the ovule containing the egg, fertilizes the egg. Plants with this type of seed and reproduction are all called 'angiosperms' and include all plants which have true flowers. They are so highly organized it is impossible to decide which is the simplest form of flowering plant.

Reproduction in flowering plants

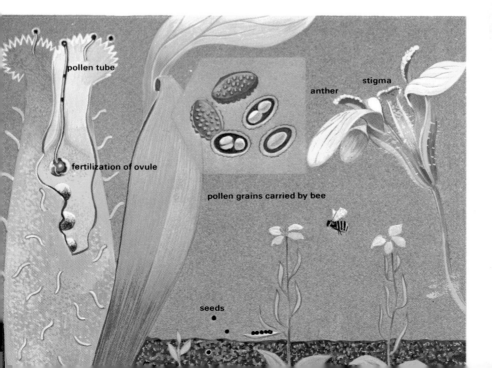

pollen tube

stigma

anther

fertilization of ovule

pollen grains carried by bee

seeds

Examples of monocotyledons

From fossil evidence it seems that angiosperms developed in the Cretaceous. Certain lines gave rise to the grasses which were very important in the evolution of the hoofed grazers such as horses, antelopes and cattle.

Several factors are evidently involved in the success of the angiosperms. Increasing aridity in late Jurassic and early Cretaceous times favored this group. Furthermore, the seas retreated, thereby exposing barren land which was quickly colonized. The conifers had depended on wind pollination and seed dispersal. However, within the angiosperms various forms of pollination and dispersal evolved, effected by insects, birds and mammals (see page 65). Although each angiosperm flower has male and female parts, there are always elaborate mechanisms to prevent self-fertilization and therefore ensure cross-pollination. Cross-pollination, as with all bisexual organisms, provides the genetic variability which is very useful to a group colonizing new areas. Many of the weed plants are angiosperms and show great adaptability in invading new habitats.

The angiosperms can be divided into two main groups, the monocotyledons and the dicotyledons. The seed contains the 'cotyledon', which is a leaf-like structure responsible for nourishing the germinating embryo. Monocotyledons have one cotyledon; dicotyledons have two.

Examples of dicotyledons

The monocotyledons of the angiosperms include many of the food crops of man, such as bananas and cereals (corn, wheat, rice, barley and oats). Also in this group are many common flowering plants, such as lilies, tulips, orchids and even palms. They have a single seedling leaf which grows to produce a plant with leaves showing parallel venation. Another distinguishing point is that no thickening of the stems takes place in the form of bark, wood or pith. Most of the monocotyledons are grass-like or small herbaceous plants, though some of the palms get fairly large.

The largest group of flowering plants are the dicotyledons. This group includes many of the broadleaf trees such as maple, oak, hickory, and the fruit trees such as apple, pear, cherry and orange. Beans, peas, tomatoes and potatoes are also dicotyledons, as are roses, asters, lilacs and many other familiar plants. The dicotyledons have double seedling leaves and yield a plant with netted or branched veined leaves.

BOOKS TO READ

The Meaning of Evolution. George Gaylord Simpson. Yale University Press, 1949.

Evolution. Life Science Library. Time-Life Books, 1962.

Evolution of the Vertebrates. Edwin H. Colbert. John Wiley & Sons, 1958.

Man, Time, and Fossils. Ruth Moore. Alfred A. Knopf, 1961.

Animal Species and Evolution. E. Mayr. Harvard University Press, 1963.

The Major Features of Evolution. George Gaylord Simpson. Columbia University Press, 1953.

Processes of Organic Evolution. G. Ledyard Stebbins. Prentice-Hall, 1966.

Evolution, Genetics, and Man. Theodosius Dobzhansky. John Wiley & Sons, 1955.

Mankind Evolving. Theodosius Dobzhansky. Yale University Press, 1962.

The Death of Adam. John C. Greene. Iowa State University Press, 1959.

Horses. George Gaylord Simpson. Anchor Books, Doubleday, 1961.

Darwin and the Beagle. Alan Moorehead. Harper & Row, 1970.

Variation and Evolution in Plants. G. L. Stebbins. Columbia University Press, 1950.

History of the Primates. W. E. LeGros Clark. University of Chicago Press, 1961.

Galapágos: The Noah's Ark of the Pacific. Irenäus Eibl-Eibesfeldt. Doubleday, 1961.

The Molecular Basis of Evolution. Christian B. Anfinsen. John Wiley & Sons, 1960.

Darwin's Finches. David Lack. Harper & Brothers, 1961.

The Origin of Races. Carleton S. Coon. Alfred A. Knopf, 1962.

Plant Life through the Ages. (2nd edition) A. C. Seward. Hafner, 1959.

The Evolution and Classification of Flowering Plants. A. Cronquist. Houghton Mifflin, 1968.

INDEX

159

OTHER TITLES IN THE SERIES

The GROSSET ALL-COLOR GUIDES provide a library of authoritative information for readers of all ages. Each comprehensive text with its specially designed illustrations yields a unique insight into a particular area of man's interests and culture.

NOW AVAILABLE

PREHISTORIC ANIMALS
BIRD BEHAVIOR
WILD CATS
FOSSIL MAN
PORCELAIN
MILITARY UNIFORMS 1686–1918
BIRDS OF PREY
FLOWER ARRANGING
MICROSCOPES & MICROSCOPIC LIFE
THE PLANT KINGDOM
ROCKETS & MISSILES
FLAGS OF THE WORLD
ATOMIC ENERGY
WEATHER & WEATHER FORECASTING
TRAINS
SAILING SHIPS & SAILING CRAFT
ELECTRONICS
MYTHS & LEGENDS OF ANCIENT GREECE
CATS, HISTORY—CARE—BREEDS
DISCOVERY OF AFRICA
HORSES & PONIES
FISHES OF THE WORLD
ASTRONOMY
SNAKES OF THE WORLD
DOGS, SELECTION—CARE—TRAINING
MAMMALS OF THE WORLD
VICTORIAN FURNITURE AND FURNISHINGS
MYTHS & LEGENDS OF ANCIENT EGYPT
COMPUTERS AT WORK
GUNS

SOON TO BE PUBLISHED